Routledge Revivals

Managing Ash from Municipal Waste Incinerators

Originally published in 1989, this report deals with issues surrounding ash residues produced by municipal waste combustors. Spurred by huge disagreements over the environmental risks that these ash residues posed; *Managing Ash from Municipal Waste Incinerators* attempts to shed light on the debates around the issue and move forward towards an appropriate solution. This title will be of interest to students of Environmental Studies.

Managing Ash from
Municipal Waste Incinerators

Alyce M. Ujihara and Michael Gough

RFF PRESS
RESOURCES FOR THE FUTURE

First published in 1989
by Resources for the Future, Inc.

This edition first published in 2016 by Routledge
2 Park Square, Milton Park, Abingdon, Oxon, OX14 4RN
and by Routledge
711 Third Avenue, New York, NY 10017

Routledge is an imprint of the Taylor & Francis Group, an informa business

© 1989 Resources for the Future, Inc.

The right of Alyce M. Ujihara and Michael Gough to be identified as authors of
this work has been asserted by them in accordance with sections 77 and 78 of
the Copyright, Designs and Patents Act 1988.

Publisher's Note
The publisher has gone to great lengths to ensure the quality of this reprint but
points out that some imperfections in the original copies may be apparent.

Disclaimer
The publisher has made every effort to trace copyright holders and welcomes
correspondence from those they have been unable to contact.

A Library of Congress record exists under LC control number: 90169011

ISBN 13: 978-1-138-95663-6 (hbk)
ISBN 13: 978-1-315-66171-1 (ebk)
ISBN 13: 978-1-138-95732-9 (pbk)

Managing Ash From Municipal Waste Incinerators

A Report

Alyce M. Ujihara
Michael Gough

November 1989

Center for Risk Management
Resources for the Future
1616 P Street, NW
Washington, D.C.

About RFF

Resources for the Future (RFF) is an independent nonprofit organization that advances research and public education in the development, conservation, and use of natural resources and in the quality of the environment. Established in 1952 with the cooperation of the Ford Foundation, it is supported by an endowment and by grants from foundations, government agencies, and corporations. Grants are accepted on the condition that RFF is solely responsible for the conduct of its research and the dissemination of its work to the public. The organization does not perform proprietary research.

RFF research is primarily social scientific, especially economic. It is concerned with the relationship of people to the natural environmental resources of land, water, and air; with the products and services derived from these basic resources; and with the effects of production and consumption on environmental quality and on human health and well–being. Grouped into four units—the Energy and Natural Resources Division, the Quality of the Environment Division, the National Center for Food and Agricultural Policy, and the Center for Risk Management—staff members pursue a wide variety of interests, including forest economics, natural gas policy, multiple use of public lands, mineral economics, air and water pollution, energy and national security, hazardous wastes, the economics of outer space, climate resources, and quantitative risk assessment. Resident staff members conduct most of the organization's work; a few others carry out research elsewhere under grants from RFF.

Resources for the Future takes responsibility for the selection of subjects for study and for the appointment of fellows, as well as for their freedom of inquiry. The views of RFF staff members and the interpretation and conclusions of RFF publications should not be attributed to Resources for the Future, its directors, or its officers. As an organization, RFF does not take positions on laws, policies, or events, nor does it lobby.

About the Center for Risk Management

The Center for Risk Management carries out independent research directed at improving management and control of environmental hazards while promoting economic growth and safeguarding individual liberties. Established as a part of Resources for the Future in 1987, the Center receives funding from Federal Government agencies, private foundations, and corporations.

Contents

List of Tables and Figures

Preface

This report, undertaken at the suggestion of the Advisory Council of the Center for Risk Management at Resources for the Future, is the first Center report concerned with solid waste and focuses on the ash residues produced by municipal waste combustors. The current widespread disagreement over the risks from ash has both restricted consideration of incineration as a solid waste management option and frustrated needed improvements in the way ash is managed. We are convinced that an objective look at this issue can move the controversy about ash and incineration toward a resolution.

A draft report was circulated for review in February 1989 and was followed by a workshop in March. This workshop brought together representatives from government, industry and environmental organizations to discuss the issues examined here. A list of reviewers and workshop participants is included at the end of this report. We are indebted to these individuals for the many valuable comments and suggestions they provided. We are also grateful to Elaine M. Koerner for her editorial and writing suggestions, John Mankin for his secretarial support and Suzanne Lewis for help in producing the final copy. We especially thank Marilyn Voigt for her assistance and support throughout the preparation of this report.

Alyce M. Ujihara
Michael Gough
Washington, D.C.   Center for Risk Management
November 1989   Resources for the Future

Executive Summary

FINDINGS

Decisions on how to appropriately manage municipal waste combustor (MWC) ash remain clouded with confusion and controversy. Federal laws and regulations that mandate how ash is to be managed are unclear, and tests for determining whether it is a hazardous waste produce ambiguous results. This situation has served both to restrict the use of incineration and to delay federal action to improve ash management. Although decisions about solid waste management, including the acceptability of incinerators, should be made locally, the federal government should not eliminate or restrict any solid waste management option—such as incineration—unless it has evidence that the option presents an unreasonable or unmanageable risk to human health or the environment. And although MWC ash contains toxic chemicals that, under certain worst-case conditions, could contaminate groundwater supplies, pollution problems from ash can be prevented by proper management.

The controversy surrounding ash brings to light some of the inherent problems with our current system for managing this nation's wastes. Dividing all wastes into but two categories—hazardous and nonhazardous—for purposes of regulation provides few options for wastes such as MWC ash which do not fit clearly into either. Another problem is that because the system focuses on how to manage the waste after it is produced, there are few incentives to use innovative treatment methods or source reduction measures. A risk–based system would improve the management of ash and other wastes. That system, however, must await the development of appropriate tests and policies.

RECOMMENDATIONS

Congress needs to enact legislation that outlines a system for managing ash.

Specifically, legislation now being considered should determine whether or not ash is to be classified as a hazardous waste. This is the single most important action that could be taken to improve the management of MWC ash. Lack of clarification on this issue burdens local and state governments that must make decisions about incinerators and allows some unacceptable management practices to persist. Without legislation, EPA almost certainly will not address the ash issue, and decisions in individual court cases, whatever the outcome, will not resolve underlying issues.

While waiting for congressional action, EPA should provide technical guidance about managing ash, pursue test development, and collect field data.

Technical guidance would provide needed leadership to state and local governments. In addition, EPA is the most appropriate agency to develop better laboratory tests to characterize ash and to collect field data. In particular, field data could support future modifications to ash regulations and provide information for validating tests.

Until more information is gathered about actual field leachates, ash should be disposed of in monofills (landfills that accept only ash) that have at least two liners, a leachate collection system, and groundwater monitoring.

Monofills (with either a single composite liner or two synthetic liners) are the most sensible short-term management approach for two reasons. First, laboratory test data show that acidic conditions, which can be caused by municipal solid waste, promote the leaching of toxic constituents—particularly lead and cadmium—from the ash. Second, ash disposed of in monofills is recoverable, unlike that which is mingled with municipal solid waste. Although monofill disposal is preferred, some codisposal of ash and municipal solid waste might be allowed. Codisposal of ash with noncombustible or nonleachable items, which should have no effect on pH and contribute little to toxicity, could be allowed under the same containment as required for monofills. Codisposal of ash and municipal solid waste, on the other hand, should be allowed only with more stringent containment.

Not all ash should be classified as a hazardous waste.

If all ash is classified as hazardous there is little incentive to reduce ash toxicity through treatment or source reduction measures. Use of these measures would be encouraged by development of appropriate tests for characterizing ash and more options for managing ash.

In the long term, the management of MWC ash should be based, at least in part, on risk assessment derived through testing.

As tests are improved, they should be used as the basis for the ash management system. This move would encourage source reduction and treatment efforts because incinerator owners or operators that use tests could use less stringent (and less costly) containment methods. Two approaches to testing would be to mimic long-term conditions at monofills and at codisposal landfills.

Research should be continued to improve our knowledge about ash and our ability to estimate risks.

The last chapter of this report contains recommendations directed at more specific problems in test development, perfecting source reduction, and dealing with particular problems associated with alkaline ash and certain physical properties of ash.

THE ASH CONTROVERSY

As the volume of municipal solid waste increases, and as existing landfill capacity diminishes, local governments must find new ways to manage their garbage. Their options are limited to source reduction, recycling, incineration using MWCs, further landfilling, or some combination of these strategies. The capacity of MWCs to reduce the volume of wastes that require disposal makes them attractive to many local governments. Already about one-tenth of the U.S municipal solid waste is being incinerated, and this proportion is projected to rise sharply in the next decade. However, the rate of increase is impossible to project, partly because incineration faces intense public scrutiny and opposition due to its perceived adverse health effects. Engineering modifications to MWCs have reduced concerns about the possible health risks from air emissions, and therefore attention has now shifted to the ash residues that remain after incineration. When garbage is burned, certain toxic constituents, particularly metals, are concentrated in the ash. If not properly managed, these constituents can leach from ash deposited in landfills and contaminate groundwater or surface water supplies.

One source of the controversy over MWC ash is its uncertain regulatory status. Under the Resource Conservation and Recovery Act (RCRA), all wastes are categorized as either hazardous (regulated under Subtitle C of RCRA) or nonhazardous (under Subtitle D). Wastes classified as hazardous must be managed under much more stringent and costly requirements than nonhazardous wastes. Yet the law is unclear as to how ash should be disposed of and generally managed. Failure to resolve this issue has caused some existing incinerators to be temporarily shut down, hindered plans for new incinerators, and delayed improvements needed to develop a comprehensive and coherent regulatory structure for managing ash.

The confusion over the regulatory status of ash can be traced to conflicting interpretations of the household waste exclusion provision of RCRA. This exclusion states that MWCs burning only municipal solid waste are not managing hazardous waste, yet it says nothing about the ash residues that remain in incinerators after combustion and which must be disposed of. Two interpretations have been offered: one holds that MWC ash is included under the household waste exclusion and, thus, incinerators are completely exempt from Subtitle C; the other view holds that the resulting ash is not included under the exclusion and, therefore, if tests of ash toxicity show that it surpasses certain regulatory levels, then ash must be regulated as a hazardous waste.

The federal government has not taken a consistent stance on the regulatory status of ash—rather it has changed its position several times. Most recently, EPA stated that if ash fails the extraction procedure toxicity (EP tox) test, it must be managed as a hazardous waste. However, the agency has not enforced hazardous waste (Subtitle C) requirements on MWCs.

Amid this lack of federal clarity on the regulatory status, some states have moved forward with their own decisions about ash management. Not surprisingly, even though states base their requirements on RCRA, widely different approaches have been taken. Some states do not apply the household waste exclusion to ash and thus manage some ash as a hazardous waste. Other states apply the household waste exclusion and exempt all ash from Subtitle C requirements despite test results. Several states that apply the exclusion have developed specific requirements under which ash is regulated much more stringently than municipal solid waste, whereas others manage ash no differently from municipal solid wastes and, in fact, allow codisposal of ash and garbage.

TOXICITY TESTING OF ASH

The analysis of MWC ash through laboratory tests and field data provides some insight into the likelihood of toxic substances leaching from the ash and causing harm to human health and the environment. Chemical analyses provide information about the identity and concentrations of substances in ash, but they provide little insight into what concentrations can become soluble and leach to groundwater or surface water. Studies of laboratory tests do show, however, that lead and cadmium—the toxic substances of greatest concern in ash—tend to be more concentrated in the fly ash than in the bottom ash. (Fly ash consists of the airborne particles that are captured by filters in the incinerator stack. Bottom ash is the heavy residue found at the bottom of the incinerator after the municipal solid waste is burned). Most MWCs in this country combine fly ash and bottom ash within the facility to produce combined ash.

The EP tox test is one method EPA uses to decide whether a waste is hazardous, and it has played a central role in the debate over MWC ash. The test was designed to mimic what could happen to a waste under worst–case disposal conditions, that is, if ash were disposed of in a municipal solid waste landfill where acids from decaying garbage could mobilize toxic chemicals. The test is conducted by mixing several samples of the waste with an acidic extraction fluid and then analyzing the resulting solution, or leachate for 14 toxic constituents. Next, the individual results are combined to obtain summary values. If any one of the toxic constituents of these summary values is above a certain concentration, the waste is deemed hazardous. When MWC ash is subjected to this test, some ash fails because of elevated levels of lead, cadmium, or both. (The current EP tox limits for lead and cadmium are 5 and 1 parts per million (ppm) respectively.)

Results of EP tox tests of ash samples can be generalized as follows: almost all fly ash samples exceed EP tox limits for both lead and cadmium;

most combined ash samples surpass EP tox limits for lead, but not for cadmium; and some bottom ash samples are below the limit for lead and all are below the limit for cadmium.

Caveats must be attached even to these generalizations, however, because of inherent problems with testing ash. One problem is that ash is very heterogeneous and thus there are no good ways to ensure that any sample is representative of the ash from an incinerator. Another problem is that results from analyses of the same sample can vary from laboratory to laboratory. At least part of this variation is due to subtle differences in test procedures that influence the solubility of lead. In particular, minor differences in the acidity of the extraction solution can produce widely differing concentrations of lead in the leachates.

EPA plans to simplify test procedures by replacing the EP tox test with a new test called the toxicity characteristic leaching procedure (TCLP). This move will eliminate some of the problems associated with the EP tox test. At present, however, it is difficult to draw conclusions about whether TCLP results will differ from those from the EP tox test because there have been few comparative analyses on MWC ash.

The chemical nature of some fly ash generates another specific concern about leachates. Because lime is added to the stack to control acid gas emissions, MWCs that use scrubbers produce highly alkaline ash. When alkaline ash is tested using only water as the extraction fluid, the leachates consistently contain lead in concentrations that exceed EP tox levels. The regulatory impact of such results is not clear.

Despite the great interest in risks from ash, there are few analyses of the leachates from actual landfills where ash has been placed. Indeed, there is often no regulatory reason to collect such data because classification of a waste as hazardous or nonhazardous depends on the results of the EP tox test or other laboratory tests, not on measurements of actual leachates. Limited field data do reveal that concentrations of lead and cadmium in leachates from "real–world" ash disposal sites are generally below EP tox limits, but a few samples have approached the limit for lead. However, field data have been collected for only a few years and there are no data on field leachates for longer periods of time.

TREATMENT AND REUSE

Eventually, all MWC ash will either be disposed of in landfills or reused in one form or another. Past hazardous and solid waste management practices have emphasized methods to contain the wastes and their leachates at landfills. This same approach, with a focus on the number and type of liners and related engineering options, has carried over to the management of MWC ash.

Treating MWC ash to prevent toxic substances from leaching out is an alternative approach. Source reduction—the removal of the toxic substances from the waste either before or after incineration—is the most direct treatment method. More common treatment options involve solidification or stabilization of the ash to impede leaching.

Some U.S. companies already offer treatment methods designed to reduce the leachability of toxic substances from ash. Spokespersons for these firms claim that treated ash passes the EP tox test and, therefore, could be safely disposed of in a nonhazardous waste landfill. Nevertheless, treatment is not widely used because it adds some costs to ash disposal and regulatory agencies do not currently require it.

The physical and chemical properties of MWC ash would allow its use in roadbed materials and in the fabrication of building blocks and concrete, among other things. The eventual market for reusing ash depends on the degree to which residual concerns about its risk can be allayed and, to a lesser extent, on the validation by regulatory agencies of treatment methods.

POSSIBLE OUTCOMES OF THE ASH CONTROVERSY

A combination of court decisions and legislative and regulatory actions will dictate a resolution to the ash problem. Although Congress is likely to decide the issue eventually, court cases brought by an environmental organization against two MWCs could help clarify the regulatory status of ash in the meantime. The two facilities produced ash that surpassed EP tox test limits but because of a regulatory exemption (the household waste exclusion), the owners/operators decided not to manage their ash as a hazardous waste. A judicial decision against the MWCs would force EPA to rely upon the EP tox test for determining the regulatory status of ash. Because many ash samples surpass the EP tox test regulatory limits, such a development, which would require regulation of at least some ash as a hazardous waste, could spur Congress on to enact legislation.

In 1988, EPA asked Congress to resolve the controversy by clarifying the regulatory status of ash. Two bills have been introduced this session but neither has passed. As with past proposed legislation, both bills would declare that ash is not to be regulated as a hazardous waste but that it should be managed more stringently than other nonhazardous wastes. Legislation of this type would favor contin-

ued operation of existing MWCs and, probably, the construction of new ones.

On the other hand, all MWC ash could end up being managed as a hazardous waste in the future. If Congress does not take action at some future date, regulatory initiatives that are already in motion will lower regulatory limits for lead and cadmium in drinking water. Since the EP tox (and TCLP) limits are tied to those standards, the limits can be expected to drop also. That would mean that many more samples would routinely fail toxicity tests, and EPA might consider listing ash as a hazardous waste. If all ash were declared hazardous, its status would have a profound effect on both existing and planned incinerators and could restrict incineration of municipal solid wastes.

Another solution to the ash problem is for EPA to issue regulations or guidance in the absence of action by Congress. Such actions could be begun immediately and would provide needed federal leadership for resolving the ash controversy. But because the courts or Congress could negate EPA's efforts, it is not likely that the agency will take any action on ash under its own initiative. The current situation, in which EPA awaits a congressional decision, provides little opportunity for any organization other than Congress to make policy.

1

Introduction: Municipal Solid Waste and the Incinerator Ash Controversy

During 1988, over 160 million tons of municipal solid waste were disposed of, incinerated, or recycled in the United States; a 20 percent increase in total volume is expected by the turn of the century when the figure is expected to reach 190 million tons. More telling is the U.S. Environmental Protection Agency's (EPA's) estimate that 45 percent of existing landfills will close within the next 5 years because available capacity will be filled (U.S. EPA 1988b). In some regions of the country, no additional landfills are being planned. In other regions, additional landfills certainly will be opened, but they surely will be more expensive to site, maintain, and monitor.

In the meantime, not only are the weight and volume of waste expected to grow with the population, but per capita waste production is expected to increase as well. The proportion of paper in municipal solid waste streams increased steadily between 1970 and 1986 and is expected to continue doing so at least until the turn of the century (see table 1–1). In fact, about one–third of the weight of municipal waste comes from packaging materials, and about half of those materials are paper (McCarthy and Pannebaker 1988). On the other hand, the proportion of plastics increased rapidly between 1970 and 1986 but is now increasing at a reduced rate. Metals, glass, and wood have declined, probably, in part, because of their replacement by plastics in fabricating many products.

Municipal waste combustors (MWCs), also called incinerators, waste–to–energy plants, or re-source recovery facilities, reduce considerably the amount of waste that must ultimately be disposed of in landfills. However, they also generate new wastes during the combustion process that require appropriate management. Combustion products that go up the stack may be emitted into the air unless trapped and contained by air pollution control devices. The trapped airborne materials (fly ash) as well as the combustion products and noncombustible residues that remain behind in the combustion chamber (bottom ash) contain hazardous substances that require careful decisions and practices to pro-

Table 1–1. Composition of Municipal Solid Waste Stream (by weight)

Waste Components	1970 (%)	1986 (%)	2000[a] (%)
Paper	32.4	35.6	39.1
Yard wastes	20.6	20.1	19.0
Plastics	2.7	7.3	9.2
Metals	12.0	8.9	8.5
Food wastes	11.4	8.9	7.3
Glass	11.1	8.4	7.1
Wood	3.6	4.1	3.6
Rubber/leather	2.7	2.8	2.3
Textiles	1.8	2.0	2.0
Other	1.7	1.8	1.9

Source: James E. McCarthy and Renee E. Pannebaker, 1988. "Solid Waste Management." Issue Brief, Environment and Natural Resources Policy Division, Congressional Research Service (Washington, D.C.: Library of Congress, Aug. 2).

[a] Projected

tect human health and the environment. The merits and drawbacks of MWCs are being considered by many local governments in the United States faced with growing solid waste problems.

Source reduction and recycling are two other options that can reduce the amount of waste that must be landfilled. Source reduction—efforts to reduce the amount of waste generated in the first place—has much potential for reducing the solid waste problem, but making changes in production methods and consumer preferences will take time. On the other hand, recycling can be put into effect in a shorter period of time. Currently about 10 percent of municipal waste is recycled. Newspaper, bottles, and aluminum cans are recycled with varying degrees of success in mandatory and voluntary programs in various cities and states, and there is general agreement that more recycling is possible. According to some experts, as much as 50 percent of solid wastes in the United States could be recycled, but achievement of such high rates would require a complete change in the way municipal wastes are currently managed (McCarthy and Pannebaker 1988). EPA and the Environmental Defense Fund have described successful recycling programs (U.S. EPA 1989; EDF 1988) and those references are suggested to readers. Other countries, particularly Japan, provide examples of what this potential might be. Although the exact rate of recycling in Japan is uncertain—estimates range from 10 percent (Shaub 1988b) to 50 percent—the best estimate is probably between 30 and 35 percent (Shaub 1988c).

Recycling and incineration are not necessarily either/or choices. For instance, several counties in the District of Columbia metropolitan area are considering construction of MWCs along with recycling programs. The appropriate combination of these complements to landfilling will probably vary with locality, but as landfill volume shrinks, it is almost certain that both will be considered in most places.

CONCERN OVER INCINERATION

One hundred and fifty-five MWCs were operating in the United States in mid-1989, and this number is projected to increase to 227 by 1992 (Levy 1989). These facilities had an estimated installed capacity of 78,700 tons per day or about 16 percent of the municipal solid waste stream. If the anticipated rate of MWC construction is realized, EPA estimates that about 28 percent of municipal solid wastes will be incinerated by 1992. Because incineration reduces the volume of municipal waste by up to 90 percent, achievement of this level of incineration would considerably reduce the demand for landfills. Furthermore, burning *all* municipal solid waste in the United States could theoretically produce electricity equivalent to the burning of 600,000 barrels of oil per day, or about 4.5 percent of the electricity generated in 1986 (McCarthy and Pannebaker 1988).

Despite these advantages, proposals for construction of new MWCs are always opposed. A primary reason for opposition to MWCs is concern about health risks from airborne emissions and ash. EPA estimated that inhalation of airborne toxic chemicals from existing and proposed MWCs would be associated with between 5 and 60 cases of cancer per year, and that the addition of devices for further control of airborne emissions would reduce those risks to between 0.5 and 4 cancer cases annually (U.S. EPA 1987). In April 1988 the former EPA Administrator Lee M. Thomas summarized the efforts of his agency in dealing with airborne emissions:

> with the air pollution control guidance issued by EPA in June, 1987 and the regulations currently being developed and expected to be proposed in November, 1989, I believe MWC air emissions can be safely controlled (Thomas 1988, p. 7).

As concern about airborne emissions has decreased, attention has become focused on risks from some metals and some organic chemicals in incinerator ash. Metals are not destroyed by incineration; in fact, they are more concentrated in fly ash and bottom ash than in the trash that enters the incinerator. The concentration is simply the result of destruction of burnable materials that leaves the metals behind. Although most organic substances are destroyed by incineration, some toxic substances—namely dioxins and their close relatives, furans—are formed during the combustion process. And, although efforts are being made to find uses for ash in construction materials, the vast majority of MWC ash continues to end up in landfills. Therefore, concerns about ash center on risks that can arise from pollution of groundwater supplies from landfills containing ash.

There is also concern about worker exposures to ash in MWC plants and at landfills and about risks to the public from windblown ash dust, particularly during transportation of ash from MWCs to landfills. Shaub (1988a) reviewed the available information about such risks, including a 1982 National Institute for Occupational Safety and Health report that concluded that toxic metals were "either not detectable or were detected only at negligible concentrations" in MWC facilities. On the basis of the limited information currently available, Shaub concludes that risks to the public from ash dust are quite low. A study by Kellermeyer (1989) reached the same conclusion. Although it is expected that continued attention will be given to such exposures, they do not appear to be as serious as possible exposures from contamination of water supplies by landfills.

CURRENT SCHEME FOR ASH MANAGEMENT

Wastes that exceed regulatory limits for being ignitable, corrosive, reactive, or toxic are defined as hazardous, unless specifically exempted, and they are regulated as hazardous wastes. Municipal solid wastes are excluded from being classified as a hazardous waste, but there is a dispute over whether the ash that remains after incineration is also exempt. Ash has none of the first three characteristics of a hazardous waste, but when EPA's standard test for toxicity is applied to ash, some samples exceed the regulatory limit. In the current dichotomous waste classification scheme—hazardous or nonhazardous—some ash would be classified as the former, some the latter.

The possible classification of ash as a hazardous waste is a critical issue. It would impose higher costs on the disposal of municipal solid waste because the costs of construction and maintenance of a hazardous waste disposal facility are greater than those for nonhazardous facilities. Also, there are many fewer hazardous waste landfills than municipal waste landfills, and so ash would often be transported further at greater cost. In addition, the potential liability incurred by users, owners, and operators of hazardous waste sites is greater than that for users, owners, and operators of nonhazardous waste disposal sites.

At present, RCRA regulations do not explicitly require that MWC ash be tested for toxic properties, and most is being disposed of as nonhazardous waste. Because there are no federal requirements specifically for ash management, most ash, by default, is managed under the existing weak requirements for Subtitle D (nonhazardous) landfills, which contains no design or performance specifications. EPA issued more rigorous draft regulations for nonhazardous waste (Subtitle D) landfills in August 1988, but the final version of these requirements is not expected to go into effect until at least 1991.

PROPERTIES OF ASH

Through the extraction procedure toxicity (EP tox) test, EPA has established maximum allowable concentrations for certain toxic constituents that can be liberated from wastes under the mildly acidic conditions that prevail in municipal solid waste landfills. Although EP tox tests of MWC ash reveal that two toxic metals—cadmium and lead—are often liberated in concentrations sufficiently high to classify the ash as a hazardous waste, there is great disagreement about the role of the test in dictating how ash should be managed.

The interpretation of results from the EP tox test is not straightforward. One major problem is that the reproducibility of test results is poor. In addition, although conditions in a landfill have a pronounced effect on the type and quantity of toxic chemicals that might be released from it, the test does not adequately mimic actual disposal conditions either at monofills (where only ash is disposed) or at codisposal facilities (where ash and municipal solid wastes are disposed of together). Other tests have been developed, but their usefulness has not been demonstrated and none has been accepted by EPA. Measurements of concentrations of metals in the liquids (leachates) that come out of operating landfills would provide straightforward information for estimating risk, but few such measurements are available.

Despite these indications that the EP tox test may not always be reliable, the federal government continues to rely heavily on this test to determine whether or not wastes should be classified as hazardous on the basis of toxicity. The future use of the EP tox test or its replacement, the toxicity characteristic leaching procedure (TCLP), in ash management almost certainly depends much more on the results of lawsuits, congressional action, or EPA decisions than on technical determinations of whether the test is predictive.

PENDING POLICY DECISIONS

Lawsuits have been brought against two incinerator operators to force them to dispose of ash that flunks the EP tox test as a hazardous waste. If those lawsuits are successful, it can be anticipated that EPA would have to require that all operators test their ash and dispose of any that exceeds the EP tox test limits as a hazardous waste. If the agency fails to do so, it will almost certainly be sued to force it to require testing and enforcement of Subtitle C rules. Hearings about the lawsuits have not yet been scheduled, but depositions are complete in one case and under way in the second.

EPA produced "Draft Guidance: Municipal Waste Combustion Ash" in March 1988 but never officially released it (U.S. EPA 1988a). That document provided technical guidance about landfill design and operation but ignored whether ash should be regulated under Subtitle C or Subtitle D. The development of that draft illustrates that EPA can provide technical guidance even when legal questions remain unsettled. Although it prefers that Congress

clarify the legal issues before it acts on the matter of ash, EPA could go ahead with technical advice or even regulation in the absence of or in anticipation of congressional action.

Meanwhile, Congress is considering legislation that would regulate MWC ash under Subtitle D. Enactment of any of these bills would eliminate use of the EP tox test and the dichotomous classification of ash as hazardous or nonhazardous. It would also direct EPA to develop regulations for the disposal of ash, and, in essence, make ash a "special" waste, with specified disposal requirements. Such legislation was not passed during the 100th Congress, but two bills have been introduced in the 101st Congress.

CONTENTS OF THIS REPORT

The remainder of this report is divided into five chapters. Chapter 2 discusses current federal regulations governing the disposal of hazardous and nonhazardous wastes and how these apply to ash. It also presents a survey of state requirements for managing ash. Chapter 3, the major technical focus of this report, discusses methods available for testing ash, an analysis of the results of those tests, some of the problems with those tests, and the limited information available from actual field measurements of leachates from ash disposal sites. Chapter 4 describes options for ash landfills and for treatment and reuse of ash. Chapter 5 discusses possible solutions to the dilemmas presented by MWC ash. Legal, legislative, and regulatory solutions are considered, and the effects of possible solutions on the future of MWCs are projected. The final chapter contains specific short–term and long–term recommendations for the management of ash under Subtitle D regulations. That chapter also describes areas of further research on ash.

REFERENCES

EDF (Environmental Defense Fund). 1988. *Coming Full Circle: Successful Recycling Today.* (New York, N.Y.: EDF).

Kellermeyer, David A. 1989. "Quantitative Health Risk Assessment for Incinerator Ash Disposal." Paper presented at MSW Incineration Ash Conference, Orlando, Fla., December 1, 1988, HDR Engineering, Inc.

Levy, Steven J. 1989. "28% of Waste Stream to be Burned by 1992, Says the EPA," Letters to the Editor, *Waste Age* vol. 20, no. 8 (August), p. 42.

McCarthy, James E. and Renee E. Pannebaker. 1988. "Solid Waste Management." Issue Brief, Environment and Natural Resources Policy Division, Congressional Research Service (Washington, D.C.: Library of Congress, August 2).

Shaub, W. M. 1988a. "Incineration—Some Environmental Perspectives." Submitted for the written record at the Office of Technology Assessment, Oceans and Environment Program, Workshop on Incineration/Waste–to–Energy Issues, June 28, 1988.

Shaub, W. M. 1988b. "More About Recycling in Japan." *CORRE Newsletter*, Coalition on Resource Recovery and the Environment, vol. 2, no. 11, p. 1.

Shaub, W. M. 1988c. "Office of Technology Assessment Project Director Questions 50 Percent Recycling in Japan." *CORRE Newsletter*, Coalition on Resource Recovery and the Environment, vol. 2, no. 9, p. 1.

Thomas, L. M. 1988. Statement before the Subcommittee on Transportation, Tourism, and Hazardous Materials of the House Committee on Energy and Commerce, April 13.

U.S. EPA (U.S. Environmental Protection Agency). 1987. *Municipal Waste Combustion Study: Report to Congress* (Washington, D.C.: EPA/530–SW–87–021A).

U.S. EPA (U.S. Environmental Protection Agency). 1988a. "Draft Guidance: Municipal Waste Combustion Ash." OS-WER Policy Directive No. 9573.00.1 (Washington, D.C.: EPA/530–SW–88–006, March).

U.S. EPA (U.S. Environmental Protection Agency). 1988b. *EPA Report to Congress—Solid Waste Disposal in the United States—Volume I* (Washington, D.C.: EPA/530–SW–88–011, October).

U.S. EPA (Environmental Protection Agency). 1989. *Recycling Works! State and Local Solutions to Solid Waste Management Problems.* (Washington, D.C.: EPA/530–SW–89–014, October).

2

Requirements for Managing Hazardous and Nonhazardous Wastes

A combination of federal and state requirements dictates the management of MWC ash. At the federal level, the Resource Conservation and Recovery Act (RCRA) of 1976 divides all solid wastes into two categories for purposes of regulation: those that are hazardous and those that are not. However, some wastes—including municipal solid wastes—are exempted from being classified as hazardous.

Much of the current controversy about ash management arises because ash—unlike most other wastes—does not fit clearly into one or the other category. Based on its properties, some ash would be classified as hazardous, some would not. Furthermore, although municipal solid wastes are exempted from being classified as hazardous wastes, it is not clear whether the exemption extends to the ash residues that result from burning such waste. Complicating the issue even more, in the absence of a clear federal position, many states have acted on their own and have taken a variety of approaches to address the ash issue.

SUBTITLE C: HAZARDOUS WASTE

EPA is responsible for implementing RCRA. Subtitle C of RCRA lays out strict requirements for managing hazardous wastes. Depending on how one interprets RCRA's legislative history and EPA's regulatory policy, MWC ash may or may not be exempt from the Subtitle C hazardous waste requirements. Until this issue is resolved, there is

little incentive for owners/operators of incinerators to test the ash they produce, and without such test results there is no clear basis for categorizing ash as a hazardous waste. However, when samples of incinerator ash have been tested using EPA's extraction procedure toxicity (EP tox) test, the results show that some samples could be classified as hazardous and therefore subject to Subtitle C regulations. Unless ash is specifically exempted from Subtitle C—and this point remains unresolved—at least some of it could be regulated as a hazardous waste.

Subtitle C Requirements

Subtitle C requirements are designed to ensure that hazardous wastes will be properly managed from the waste's generation to its final disposal. Generators of hazardous waste must comply with certain waste identification and record-keeping requirements. Transporters of hazardous waste must meet U.S. Department of Transportation rules that include requirements for labeling, using proper containers, and reporting spills. Generators as well as transporters must also obey manifest requirements which are designed to track the waste from its generation to its final disposal.

Requirements for the treatment, storage, and disposal of hazardous waste are the most detailed part of Subtitle C. The owner/operator of a new treatment, storage, or disposal facility must have a permit to operate. In addition, existing land disposal facilities (including landfills) would have had to have obtained a permit by November 1988 to continue operation. All facilities must meet specific design and operating requirements to receive a permit. One of the most important elements of the facility's design is the liner and leachate collection system, which must be designed to prevent movement of toxic substances from the landfill into groundwater or surface water supplies. Most facilities are also required to have monitoring wells that reach into the uppermost aquifer underlying the facility. The water from those wells must be monitored for certain hazardous constituents. In general, acceptable concentrations for hazardous constituents in monitoring wells are equivalent to drinking water standards. If the standard is violated, the owner/operator must conduct a corrective action—a program to remove or treat the contaminated groundwater until the standards are no longer exceeded.

Design standards for "existing" landfill facilities built before the amendment of RCRA by the Hazardous and Solid Waste Amendments (HSWA) in late 1984 are less stringent than standards for "new" facilities built after that date. For example, existing facilities need only a single liner, whereas new hazardous waste landfills must have two or more liners with a leachate collection system above and between the liners. The double-liner requirement, however, may be waived for certain types of waste, such as that from foundry furnace emission controls or metal-casting molding sands—though that waste must be placed in a monofill with at least one liner. (Landfill liner systems are described in Appendix 2–A to this chapter.) Furthermore, requirements can be waived if the owner/operator demonstrates that an alternate system performs as well as a double-liner system.

Between 1976 (when RCRA was exacted) and 1984 (when it was amended) hazardous waste disposal most often meant containment of the waste in a well-lined and monitored facility. Since 1984, however, EPA has begun to move away from direct land disposal of untreated hazardous wastes. HSWA requires that all hazardous wastes,

unless exempted, will either be banned from land disposal or at least undergo some form of treatment prior to disposal.

Subtitle C further requires that the owner/operator of a disposal facility develop a plan to close the facility and to take care of the facility for at least 30 years after closure ("post-closure care"). Owners/operators must also prove that they have the financial resources necessary for closing and conducting post-closure care activities at the facility. Finally, people who violate these rules can face civil and criminal penalties.

Defining a Hazardous Waste

There are several criteria by which a solid waste can be defined as hazardous under RCRA.

1. Wastes that are tested and shown to exhibit certain hazardous characteristics are defined as hazardous wastes. Under existing regulations, these characteristics include ignitability, corrosivity, reactivity, and extraction procedure (EP) toxicity. The EP toxicity characteristic refers to the extent to which toxic constituents can be extracted from a waste under certain specified conditions.

2. Wastes that are known to contain any of a number of toxic constituents listed in the *Code of Federal Regulations* (40 CFR Part 261, Subpart D, Appendix VIII), may also be defined as hazardous wastes on the basis of the EPA Administrator's determination.

3. Wastes can be defined as hazardous if they are shown to be acutely toxic to humans on the basis of human or animal studies.

4. Entire mixtures that are known to contain hazardous wastes may be judged hazardous.

Wastes meeting the criteria for hazardous waste generally are placed on a "List of Hazardous Wastes" (40 CFR Part 261, Subpart D). People may petition EPA to have a waste taken off this list, which is called "delisting." To do so, however, the petitioners must demonstrate that in general the waste no longer exhibits properties for which it was judged to be hazardous. For instance, a petitioner could show that treatment of the waste had changed the properties of the waste.

Exemptions from Subtitle C

Some wastes are exempt from regulation under Subtitle C even though they may exhibit characteristics that would otherwise define them as hazardous. Among the exempted materials are agricultural wastes, mining wastes, ash from coal or other fossil–fuel combustion, and small–quantity generator wastes (from operations that generate less than 100 kilograms of waste per month). In addition, municipal solid wastes are exempt from regulation under Subtitle C. The household waste exclusion, which applies to MWCs burning municipal solid waste, is the source of much confusion and controversy about ash.

Household Waste Exclusion and Its Connection to Ash

The household waste exclusion, which could exempt most MWCs from Subtitle C requirements, plays a major role in the unresolved controversy over the proper regulatory treatment of MWC ash. Originally, the exclusion (which first appeared with EPA's hazardous waste regulations promulgated on May 19, 1980, under RCRA) applied to facilities that burned only wastes from households, and the exemption was lost if the household wastes were mixed with any nonhazardous commercial or industrial wastes. Facilities that fell under the exclusion were exempt from the Subtitle C requirements even if their wastes exhibited hazardous characteristics. The reason for so liberal an exclusion was that EPA assumed that neither household wastes nor the ash produced from these wastes could be expected to exhibit any hazardous characteristics because of the small quantities of hazardous materials present in the original waste (U.S. EPA 1983). Furthermore, although EPA's regulations specifically address the *waste* from households, and not the resultant incinerator *ash*, EPA's operating policy—as de-

scribed in the preamble to the 1980 regulations—was to apply the household waste exclusion to the entire waste stream, which includes incinerator ash (Kovacs 1987).

Congress attempted to clarify and expand the household waste exclusion under the HSWA in 1984. Section 3001(i) of HSWA, entitled "Clarification of the Household Waste Exclusion," states that all resource recovery facilities (incinerators that recover energy) "shall not be deemed to be . . . managing hazardous waste" if they burn only municipal solid waste. This waste, of course, includes nonhazardous commercial and industrial wastes in addition to household wastes and by applying the exclusion to facilities burning municipal solid waste, Congress greatly expanded its scope. Unfortunately, the clarification, as with the original exclusion, only specifically discusses *wastes*. It does not mention *ash*. This wording has caused confusion because it is not clear whether facilities that burn municipal solid waste yet produce ash judged to be a hazardous waste are still exempt under Section 3001(i). Presumably, Congress wanted to encourage energy recovery at incinerator facilities when it stipulated the Section 3001(i) exemption requirements because incinerators that *do not recover energy* are not exempt. Ash from such incinerators could be subject to Subtitle C regulation if it fails the EP tox test.

In 1985, EPA shifted its interpretation of Section 3001(i) in a way that deviated from the agency's past policy toward incinerator ash. EPA stated in the preamble to the rule codifying the 1984 HSWA that it did *not* intend to exempt ash residues from waste–to–energy facilities if the ash routinely exhibited hazardous characteristics (U.S. EPA 1985). Note that when EPA took this position in 1985 the agency also had "no evidence to indicate that these [MWC] ash residues are hazardous under existing rules" (U.S. EPA 1985).

However, late in 1987, EPA once again changed its position and backed away from its 1985 interpretation. EPA's Assistant Administrator J. Winston Porter, in testimony to Congress, conceded that EPA's 1985 interpretation that ash residues could fall under Subtitle C "may have been in error" (Porter 1987). That same month it was reported that EPA would publish a *Federal Register* notice to resolve the ash issue by clarifying the household waste exclusion. That notice was to be supported by a legal determination from EPA's Office of General Counsel (BNA 1987). EPA's plans changed, however, and the agency has not yet published any official clarification of the household waste exclusion. In March 1988, EPA Administrator Thomas stated that the agency had decided to delay issuing a policy decision (Thomas 1988) while it awaited clarification from Congress (BNA 1988b). More recently, in testimony before Congress in May 1989, EPA appears to have reverted to its 1985 interpretation—that ash could be a hazardous waste if it exhibits a hazardous characteristic (Lowrance 1989). It has not, however, taken any actions to enforce this new stance. A summary of EPA's apparently changing positions on the household waste exclusion is presented in table 2–1.

In sum, it is not clear whether ash from municipal solid waste burned in MWCs is subject to Subtitle C requirements even if it fails the EP tox test. *Waste* originating strictly from household sources is exempt, and EPA's policy since 1980 has included the *ash* from these sources as well. Few MWCs, if any, however, burn strictly household wastes. Congress allowed MWCs with energy recovery facilities to be exempted from Subtitle C if they burned municipal solid waste (household together with nonhazardous commercial and industrial wastes), though it is uncertain whether Congress meant to exclude these facilities if the ash residues they generate are judged to be hazardous. Congress may have intended to include these facilities under the household waste exclusion, but it probably did not anticipate that the ash would exhibit properties characteristic of a hazardous waste when assayed by the EP tox test.

EP Tox Test

A waste can be classified as hazardous if it is ignitable, corrosive, reactive, or toxic. The EP tox test is one means by which the agency determines whether a waste is hazardous because of its toxicity. In fact, it is the only test that could reveal a MWC ash sample to be hazardous because ash is neither ignitable, corrosive, nor reactive. The EP tox test measures the potential for a waste to liberate ("leach") hazardous constituents by mimicking conditions in a municipal solid waste landfill.

The EP tox test involves several steps. First, samples of at least 100 grams each are taken from a waste. The liquid portion of the waste is removed and any large particles are crushed so that the waste can pass through a 9.5 millimeter (0.375 inch) screen. Second, the waste sample is mixed with deionized water, and sufficient acetic acid is added to bring the pH to 5.0. The ratio of solution to waste is approximately 20 to 1 by weight. Third, the mixture is continuously agitated for 24 hours. During this period, the pH is constantly monitored and maintained at 5.0 plus or minus 0.2. However, there is an upper limit on the total amount of acid that can be added to maintain the pH at 5.0. Finally, after 24 hours, the mixture is filtered and the liquid portion is analyzed for the presence of 14 toxic constituents—eight metals, four insecticides, and two herbicides. The regulatory limits for the EP tox test are set 100 times greater than the drinking water standards for these 14 constituents. EPA requires that a minimum of four samples of the waste be collected over a period of time adequate to represent the waste's variability (U.S. EPA 1982, p. 5). 90 percent probability[1] the limit is determined and compared to

---

[1] This level is also called the *80 percent* confidence interval because when the samples are normally distributed, 10 percent of the samples fall outside (i.e., to the left of) the lower limit and 10 percent are outside (i.e., to the right of) the upper limit.

Table 2–1. EPA's Changing Interpretation of RCRA Household Waste Exclusion

Date	Type of Facility		
	MWCs Burning Household Waste[a]	MWCs Burning Municipal Solid Waste[b] (with energy recovery)	MWCs Burning Municipal Solid Waste[b] (without energy recovery)
May 19, 1980	These facilities are exempt from Subtitle C under 40 CFR 261.4(b); exemption applies to entire waste stream.[c]		
Dec. 15, 1983		EPA policy does not exempt these facilities from Subtitle C.[d]	EPA policy does not exempt these facilities from Subtitle C.[d]
Nov. 8, 1984		RCRA amendments exempt these facilities from Subtitle C, but ash was not addressed.[e]	RCRA amendments do not exempt these facilities from Subtitle C.[e]
July 15, 1985		Not exempt if ash "routinely exhibits a characteristic of hazardous waste."[f]	
Dec. 3, 1987		The 1985 interpretation "may have been in error...." The law was "probably intended to exclude these ash residues from regulation under Subtitle C."[g]	
Dec. 4, 1987		EPA will issue a proposal that exempts MWCs from Subtitle C.[h]	
Mar. 25, 1988		EPA will not issue a proposal; it will seek clarification from Congress.[i]	
May 11, 1989		"Our statutory interpretation of the existing law is correct," i.e., "ash...which exhibits a characteristic of a hazardous waste must be managed accordingly."[j].	

(continued)

Notes for Table 2–1.

[a] Does not include any commercial or industrial wastes.

[b] Includes nonhazardous commercial and industrial wastes.

[c] Preamble to final rule in *Federal Register* May 19, 1980, p. 33,099, as cited by William L. Kovacs, in "Can EPA Regulate Ash as a Hazardous Waste?" *Waste Age* May 1987, pp. 105–111.

[d] Letter to George W. Guhr, Solid Waste Management Board, Waukesha County, Wisconsin, from John H. Skinner, Director, Office of Solid Waste, U.S. EPA, Dec. 15, 1983.

[e] Hazardous and Solid Waste Amendments, Pub. L. 98–616, Section 3001(i), Nov. 8, 1984.

[f] Preamble to the codification of HSWA, *Federal Register*, July 15, 1985, pp. 28725–6.

[g] Testimony of J. Winston Porter, Assistant Administrator, Office of Solid Waste and Emergency Response, U.S. EPA, before the Subcommittee on Hazardous Waste and Toxic Substances of the Senate Committee on the Environment and Public Works, Dec. 3, 1987.

[h] Bureau of National Affairs, Inc., "EPA says Ash From Commercial Waste Burning Exempt from RCRA Hazardous Waste Regulations," *Environment Reporter*, Dec. 4, 1987, p. 1803.

[i] Statement by EPA Administrator Lee Thomas as reported by the Bureau of National Affairs, Inc., in "EPA Hit for Lax Superfund Enforcement Effort at House Appropriations Subcommittee Hearing," *Environment Reporter*, Mar. 25, 1988, p. 2363.

[j] Testimony of Sylvia K. Lowrance, Director, Office of Solid Waste, U.S. Environmental Protection Agency, before the Subcommittee on Transportation, Tourism, and Hazardous Materials of the House Committee on Energy and Commerce, May 11, 1989.

the regulatory threshold The probability limit establishes the concentration expected in 90 percent of all samples. If the 90 percent probability level surpasses the regulatory limit (100 times the drinking water standard) for any of the 14 constituents, the waste is classified as a hazardous waste and, unless it is exempt from regulation, it becomes subject to RCRA Subtitle C requirements. (A description of the test in the regulations can be found in 40 CFR Part 261, Appendix II.)

Rationale for Development of the EP Tox Test

The original RCRA law of 1976 defined a hazardous waste as a waste that "may pose a substantial present or potential hazard to human health or the environment when improperly . . . disposed of . . ." EPA assumed that if a hazardous waste was not managed under Subtitle C, it would most likely be disposed of in a municipal solid waste landfill. Thus, the agency used this scenario of an improperly disposed waste as the basis for the EP tox test (U.S. EPA 1980, p. 22).

The rationale used by EPA to construct the EP tox test has always been questioned. In the test, acid is added to the waste to mimic the acidic nature of an actively decomposing municipal solid waste landfill. The agency chose acetic acid as the acidifying agent because it was the most prevalent acid found in municipal landfill leachate. A pH of 5.0 was chosen because it fell "well within" a wide range of values reported in the literature for municipal waste landfills and because it promoted "relatively aggressive leaching" of constituents (U.S. EPA 1980, p. 54.) The limit on the total amount of acidity acting on the waste sample was based on field measurements and EPA estimation of the amount of acid that could potentially act on a given amount of waste.

EPA acknowledged that concentrations of toxic materials in leachate can be reduced as the toxic materials are diluted in groundwater and are precipitated out, filtered through soil, or adsorbed to components in the soil around a landfill. How-

ever, the selection of a quantitative attenuation factor—the amount of attenuation to be expected as a leachate moves through the soil and is diluted in the groundwater around a landfill—was largely arbitrary. In fact, EPA assumed that no hazardous constituents would be attenuated as the contaminated groundwater passed through the soil. The value EPA chose was based solely on the effects of dilution. Originally, EPA proposed a factor of 10 which meant that the acceptable concentrations from EP tox test results would be 10 times the drinking water standards for the 14 toxic constituents that are examined in the test. However, after much public comment, most of which argued that the level was "far too conservative," the agency reconsidered. The most important factor EPA cites for increasing the attenuation factor to 100 was "the relative absence of empirical data upon which to base an attenuation factor, and strong suggestions that choice of any of a wide range of attenuation levels could be supported by what data is [sic] available" (U.S. EPA 1980, p. 77).

Requirement for Testing

Nowhere in the RCRA regulations does it state explicitly that all waste generators must test the wastes they produce to determine whether or not their wastes are hazardous. The regulations do state, however, that if the waste is not exempt from Subtitle C and it is not already listed as a hazardous waste, then the waste generator must make a determination as to whether the waste exhibits a hazardous characteristic. This determination is made by either testing the waste or "[a]pplying knowledge of the hazard characteristic of the waste in light of the materials or the processes used" (40 CFR part 262.11(c)(2)). In other words, if an operator does not test the ash an incinerator produces, the operator makes a judgment about its toxicity based on the materials that go into the waste and the processes the waste undergoes. For example, on the basis of what is known about the composition of municipal solid waste, one could assume that it is not hazardous and not test it. (In reality, this assumption is unnecessary because municipal solid waste is already exempt from Subtitle C.)

One probably could not draw the same conclusion about the ash residues from the incineration of municipal solid waste. As EPA has stated, "residues from burning could, in theory, exhibit a characteristic of hazardous waste even if no hazardous wastes are burned, for example, if toxic metals become concentrated in the ash" (U.S. EPA 1985). Thus, if one were to ignore other factors that continue to make the status of MWC ash uncertain (i.e., some MWC ash may be exempt from Subtitle C altogether), on the basis of knowing that the process of incineration could concentrate toxics, one should probably opt to test ash from MWCs.

Changes to the EP Tox test

Toxicity Characteristic Leaching Procedure. Because of the need to expand the test to include additional chemicals and because of concern over EP tox test procedures for characterizing wastes, EPA proposed changes to the test in 1986. The new test, the toxicity characteristic leaching procedure (TCLP), will likely replace the EP tox test this year (1989). One of the main reasons for a new test was to add 38 organic chemicals to the existing 14 toxic constituents, and this change is not likely to affect the regulatory status of ash. However, changes to the test procedures will have an impact. The TCLP is very similar to the EP tox test in that it is designed to mimic the acidic conditions in a municipal solid waste landfill. The actual test procedures, however, differ. First, in the TCLP, one of two extraction fluids is added to the waste sample, depending on the results of a pretest. The EP tox test relies on a single standard extraction fluid. Furthermore, in the TCLP the acidic extraction fluid is added only at the beginning of the test, and the solution is mixed for only 18 hours. By contrast, the pH level during an EP tox test is continuously adjusted by the addition of acid as appropriate for the entire 24–hour period.

New Drinking Water Standards. Classification of ash as hazardous or nonhazardous could also be affected by proposed changes in drinking water standards, which are the basis for determining regulatory levels used in the EP tox test. EPA has proposed regulations for lowering the maximum contaminant level (MCL) for lead in drinking water from 50 parts per billion (ppb) to 5 ppb (U.S. EPA 1988a). The basis for this change is new health data that indicate that lower concentrations of lead in drinking water than previously believed can cause harmful effects.

EPA has also proposed that the new standard of 5 ppb be met at the drinking water treatment plant before the water enters the distribution system. However, because a significant source of lead in drinking water is the pipes in the distribution system, the concentration of lead "at the tap" is often greater than that which leaves the treatment plant. To address this problem, EPA has proposed allowing the public to be exposed to higher levels of lead before remedial action would be required. Thus, there are several proposed standards—a 5–ppb MCL for water entering the distribution system and higher allowable levels at the tap—and it is not obvious which level would be the basis for the regulatory limit for lead in the EP tox test (or in the TCLP, if the TCLP replaces the EP before the new lead standard becomes final).

Because the existing EP tox level is based on the MCL, it is likely that EPA will continue to rely on this standard, which would result in a 10–fold reduction to the EP tox limit. On the other hand, allowing higher levels at the tap implies that EPA regards these levels as "safe," and thus the agency could decide not to lower the EP tox test limit 10–fold to reflect the new MCL. Also possible is that EPA may choose an altogether new level for lead when the proposal becomes final. With any reduction in the EP tox limit for lead, clearly a larger proportion of ash samples would exceed the EP tox test levels.

At what point EPA will adjust the regulatory level for lead in the EP tox test is uncertain. However, it is certain that EPA will not do so before the proposal to change the MCL for drinking water becomes final. Moreover, changing the EP tox limit to match this new drinking water standard will require a separate agency ruling.

Lead is not the only drinking water constituent being examined. In May 1989, EPA proposed lowering the MCL for cadmium in drinking water (U.S. EPA 1989). The new proposed level is 5 ppb, or half the existing level. Such a change, if it is incorporated into the EP (or TCLP) test limits, would reduce that limit from 1.0 ppm to 0.5 ppm. The proposed change to the cadmium level is not as great as the proposed change for lead, but it could still have a significant impact on the regulatory status of ash. Many EP tox test measurements, for instance, of fly ash as well as combined ash are very near the possible new level of 0.5 ppm. As with lead, it is not certain that the proposed level of cadmium will be selected in the final ruling. Furthermore, EPA will not initiate changes to the EP (or TCLP) levels at least until the new drinking water standards becomes final, which will not occur before late 1990 for both lead and cadmium.

SUBTITLE D: NONHAZARDOUS WASTE

Subtitle D stipulates the regulation of nonhazardous solid wastes. It is important to discussions of MWC ash because, in the present dichotomous classification scheme for solid wastes, a waste that is not classified as hazardous and therefore not regulated under Subtitle C must by default fall under Subtitle D. Because the classification of ash is uncertain, most ash is regulated under this subtitle.

Subtitle D of RCRA also directed EPA to develop requirements for managing nonhazardous wastes. In 1979, EPA published criteria for classifying solid waste disposal facilities and disposal practices. Facilities not meeting the criteria were classified as "open dumps," which were prohibited under RCRA. These criteria are very general. Primarily, they state that a facility or practice not

contaminate a drinking water aquifer beyond the facility boundary. They also contain provisions to minimize disease and improve safety at a facility, and to limit the application of solid wastes near land used for production of food and fodder crops. These requirements do not contain design requirements or any requirements to monitor groundwater or to collect leachate. Not surprisingly, as shown in table 2–2, few currently operating municipal solid waste landfills have those features.

Table 2–2. Features of Municipal Solid Waste Landfills Operating in 1986

Features	Percent
Liners	15
Leachate collection or removal systems	5
Methane gas controls or recovery systems	17
Run–on and run–off controls	46
Waste restrictions	40
Groundwater monitoring	25
Surface water monitoring	12
Air and methane monitoring	4

Source: McCarthy, James E., and Renee E. Pannebaker. 1988. "Solid Waste Management," Issue Brief, Environment and Natural Resources Policy Division, Congressional Research Service (Washington, D.C., Updated August 2).

In the late 1970s, federal funds were available to states that developed and implemented EPA–approved solid waste management plans. However, beginning in 1981 and culminating with the enactment of HSWA in 1984, attention shifted from Subtitle D to Subtitle C and hazardous waste. Federal funds were no longer made available to states for Subtitle D activities, and there was essentially no work in this area at the federal level. In 1984, however, with the regulatory structure for managing hazardous wastes in place, Congress directed EPA to turn once more to Subtitle D and finish the task of developing requirements for managing nonhazardous wastes.

The HSWA provisions affecting Subtitle D require EPA to revise standards for municipal solid waste landfills. They also required the agency to promulgate final revisions by March 31, 1988. Although EPA did not meet that deadline, proposed criteria were released in August 1988 (U.S. EPA 1988b).

The proposed criteria included location standards for landfills, requirements for day–to–day operations, groundwater monitoring, facility closure and post–closure activities, and financial assurance. In addition, EPA proposed risk–based performance standards for the facility design requirement. This means that any combination of liners, leachate collection systems, final covers, and other features could have been used as long as they afforded the specified level of protection of human health. Because of the many comments received (BNA 1988a), EPA plans to make many substantive changes—especially to the design requirement—before the proposal is made final.

As long as ash is not regulated under Subtitle C, and barring action by Congress to regulate ash under a new set of rules, the existing Subtitle D requirements will apply to MWC ash until the agency issues its final rules (expected in early 1990). Facilities will be required to adhere to the new requirements 18 months later.

STATE REQUIREMENTS

States have taken widely different approaches for managing MWC ash and interpreting federal requirements in the absence of clear guidance. About half of the 16 states we reviewed in a recent survey (see table 2–3) base their ash management program on EPA's 1985 interpretation of the household waste exclusion. In this interpretation, EPA stated that ash could still be classified as a hazardous waste if it routinely failed the EP tox test. These states have "testing–based" systems, that is, they rely on test results to determine whether the ash is managed as a hazardous or nonhazardous waste. The EP tox test is by far the most important test for categorizing a waste as hazardous, though some states require additional tests such as total chemical analyses and water extraction tests. Several are also attempting to specify EP tox test procedures more precisely to remove some of the test result ambiguities.

Our survey revealed four states with testing–based systems that have identified some ash that fails the EP tox test and must be managed as a hazardous waste. In nearly all cases, the hazardous waste was fly ash. However, one MWC in Maine produces three waste streams: fly, combined, and bottom, and the state requires that all three be tested and managed separately. Although the combined ash passed the EP tox test, both the fly ash and the bottom ash from this facility surpassed EP tox limits for lead and thus they are being managed as hazardous wastes. In Michigan all ash (combined) from four MWCs had surpassed EP tox test limits. As a result, two of these facilities were forced to shut down; a new facility that was just starting up stored its ash under a temporary program and was in litigation; the remaining facility disposed of its ash in a substandard landfill in violation of state requirements. However, new legislation that lays out an ash management system not based on tests and mandates that ash be placed in monofills was recently enacted.

In two states with testing–based programs, no ash has yet been declared hazardous, though one such state, Oregon, notes that results are often near the regulatory threshold.

At least nine states have systems for managing ash not based on test results. These states have interpreted the household waste exclusion to apply to ash regardless of EP tox test results. Although these states may require that ash periodically be tested, in most circumstances it cannot be declared hazardous. In general, these states rely heavily on incineration for managing municipal solid waste, and, perhaps out of necessity, have developed detailed disposal requirements for ash. Typically, these requirements stipulate that ash, at a minimum, be disposed of in a monofill with a single

Table 2–3. Survey of Requirements for Managing MWC Ash in 16 States

State	Number of MWCs	Percentage of Municipal Solid Waste Incinerated	Municipal Solid Waste Incinerated (tons per day)	Testing–Based Management System	Any Ash Managed as a Hazardous Waste?	Ash Disposal Practices (for ash classified as nonhazardous)
AR	4	5	200	No	—	Codisposal.[a]
CT	5	67	5,000	No[b]	—	New regulations will require monofills with composite liner; codisposal is not allowed.
FL	10	23	9,300	No	—	Monofill and codisposal; state recommends that ash be disposed of separately from solid waste.[c,d]
MA	10	50[c]	9,000[c]	No	—	Monofill with composite liner.
MD	3	26	5,100	No	—	Codisposal with single composite liner.[a]
ME	6	73[e]	1,800[e]	Yes	Yes[f]	Monofill with composite liner plus an additional synthetic liner.[g]
MI	2	12	4,000	No[h]	—	Monofill with composite liner.[h]
MN	10(+3)	50[i]	5,500[i]	No[j]	—	Monofill with single liner or storage.[j]
NJ	2	1.5[k]	460[k]	Yes	Yes[l]	Composite liner; state recommends monofills; requirements are based on site assessment.
NY	12[m]	20[m]	10,000[m]	No[b,n]	—	Fly ash: monofill with double composite liner, combined or bottom ash, or treated fly ash: monofill with composite liner or codisposal with double composite liner.
OH	7	12	3,800	Yes	No	Segregated from municipal solid waste by soil berm; monofill with a composite liner may be required if groundwater contamination from landfill is detected.

(continued)

Table 2–3. (continued)

State	Number of MWCs	Percentage of Municipal Solid Waste Incinerated	Municipal Solid Waste Incinerated (tons per day)	Testing–Based Management System	Any Ash Managed as a Hazardous Waste?	Ash Disposal Practices (for ash classified as nonhazardous)
OR	3	12	700	Yes	No	Monofill with soil or composite liner.
PA	2	<5	700	Yes	Yes[l]	Monofill or segregated cell; liner system consists of soil layer subbase, clay or synthetic secondary liner, and synthetic primary liner.
TX	4	<1	35	No	—	Codisposal.
VA	8(+1)	10[o]	3,000[o]	Yes	No	Monofill or codisposal, with two liners (can be clay or synthetic).[a]
WI	11	8	1,300	Yes	Yes[l]	Segregated lined cell at solid waste landfill.

Notes: This survey was carried out with the assistance of the Association of State and Territorial Solid Waste Management Officials (ASTSWMO). The states surveyed here do not constitute a representative sample of all states. Instead we focused predominantly on states that presently, or will in the near future, rely heavily on incineration.

[a] Ash disposed of no differently from other municipal solid wastes.

[b] Testing of ash is required but it does not dictate how the ash is managed.

[c] New ash requirements are under development.

[d] Ash from several new facilities will go to double–lined landfills to meet local public concerns.

[e] If operating at capacity.

[f] This includes nearly all fly ash generated in the state as well as the bottom ash from one facility.

[g] Codisposal with wastes of similar pH allowed.

[h] Based on new legislation just enacted. Previously, all ash failed the EP tox test. Ash from the two operating facilities has been placed in temporary storage or substandard landfill while awaiting the new legislation. Two additional facilities had been shut down because of the disputes over the ash.

[i] This figure includes the three facilities soon to come on line.

[j] Presently, MN is managing ash under a temporary program. The state is in the process of developing a testing–based system.

Table 2–3. (continued)

k NJ has 19 facilities planned or under construction. The state plans to burn 65% (about 20,000 tons per day) of its municipal solid waste by 1993.

l Fly ash only.

m By the end of 1989.

n These requirements apply only to MWCs that have energy recovery; management of ash from facilities that do not recover energy *is* based on test results. Plus several very small nonenergy recovery facilities.

o When a 3,000 ton per day facility comes on line next year, these figures will double to 20% and 6,000 tons per day.

p Plus several very small nonenergy recovery facilities.

composite liner and a leachate collection system. Several states allow codisposal with additional containment (usually an additional liner), though at least one state (Connecticut) does not allow any codisposal. New York has developed detailed ash requirements which restrict the application of the household waste exclusion to only those MWCs that are designed to recover energy. Facilities without energy recovery are not exempt and their ash will be classified as hazardous if it fails the EP tox test. This distinction between facility type parallels RCRA, though in practice there is little inherent difference between ash from energy recovery facilities and that from nonenergy recovery facilities.

At least three states that do not have testing–based programs do not require that ash be managed any more stringently than municipal solid waste. These states allow ash to be codisposed with other garbage. Two of these states incinerate only a very small quantity of municipal solid waste and thus the incentive for instigating a special ash management program is low.

Officials from many states expressed frustration that the federal government has not resolved the ash issue. Some states are waiting for the federal government to act rather than issuing new requirements specific to ash. This lack of clarity at the federal level prompted Minnesota to enact special ash legislation directing the state to develop an entirely new program for managing MWC ash. This legislation will include a new ash leaching test that could serve as the basis for determining the level of containment needed for the ash. Until the new rules are developed, however, the Minnesota law directs the state to manage ash under an interim program, which permits ash to be placed in temporary storage or in monofills. Interim mono-

fills may or may not be made permanent depending on the new rules. The interim program terminates when the state develops new ash rules or in the event that EPA establishes testing and disposal requirements for MWC ash.

SUMMARY

The heart of the MWC ash controversy is how the ash should be regulated—as hazardous or nonhazardous waste. Regulations for hazardous wastes, of course, are much more stringent and comprehensive than those for nonhazardous wastes. The current federal regulations for the nonhazardous wastes are weak, and long overdue revisions to these requirements were finally proposed last year but have not been finalized.

Neither the law nor the existing regulatory structure provides a clear answer about how MWC ash should be categorized. Ash may be exempt altogether from being regulated as a hazardous waste even if it exhibits a characteristic of a hazardous waste, but this issue has not been resolved. If MWC ash is not exempt, the EP tox test is used to make regulatory determinations. Difficulties with that test have spurred efforts to revise it. Future reductions in drinking water standards—which are the basis for the regulatory limits in the EP tox test—could mean that most ash would fall into the hazardous waste category. A decision framework that summarizes the many factors that determine the regulatory status of ash is shown in figure 2–1.

While the stalemate at the federal level persists, states have been forced to deal directly with the ash problem. In the absence of a clear federal position, states have implemented very different management approaches.

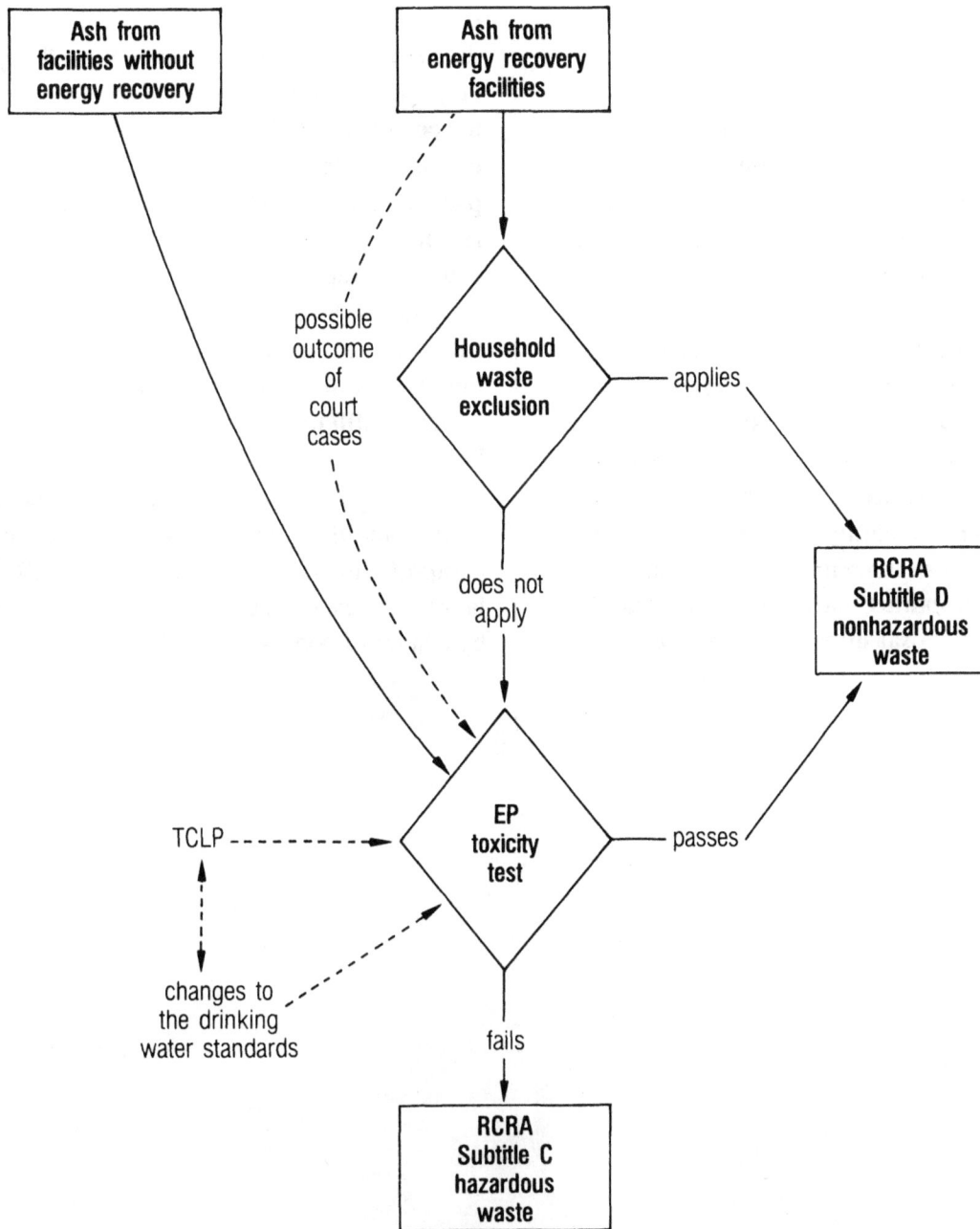

Figure 2–1. Decision Framework for Determining the Regulatory Status of Ash.

APPENDIX 2–A

Landfill Liner Systems

Landfill liners and leachate collection systems are designed to prevent or impede liquids that may be contaminated with toxic substances from moving into groundwater or surface water supplies. Figure 2–A–1 presents a schematic of several liner systems.

Flexible membrane liners are made of synthetic materials of various kinds and thicknesses. For instance, some proposed ash legislation that specifies membrane liners for ash disposal sites requires a liner at least 1/16–inch (60 mils) thickness. Low–permeability soil liners are usually compacted clay or other natural materials. Specifications for Subtitle C hazardous waste landfills, and for ash disposal in proposed legislation, require that the soil liner meet a performance standard based on the liner's permeability. This standard specifies that water cannot move through the clay liner at a rate greater than 0.0000001 centimeter per second (10^{-7} cm/sec). Composite liners consist of a flexible membrane liner above a compacted soil liner; performance and/or dimensional requirements can be specified for both components of composite liners.

The leachate collection system and drainage/filter medium are methods to collect or channel liquids from above the liners and conduct them out of the landfill to a facility for the treatment of the liquids.

A single composite liner can be combined with an additional (upper) synthetic liner and two composite liners can be combined as well. In the latter case, the two composite liners are separated by a layer of porous material.

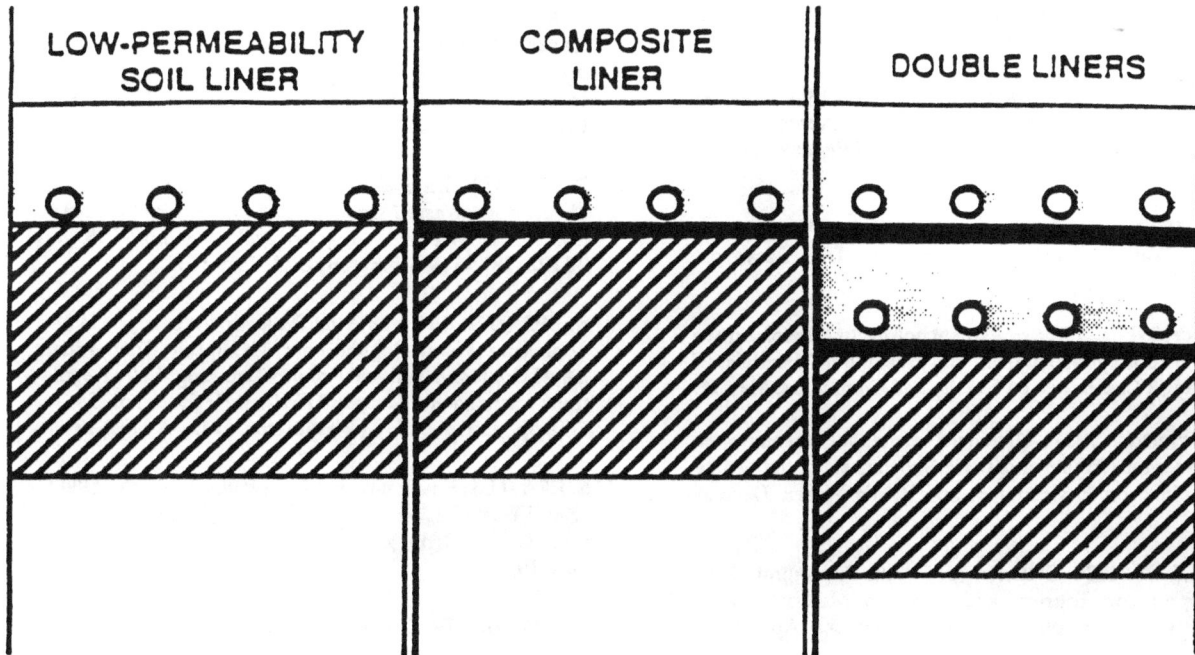

NOT TO SCALE

LEGEND:

━━━━━ FLEXIBLE MEMBRANE LINER (FML)

///// LOW-PERMEABILITY SOIL LINER

LEACHATE COLLECTION SYSTEM
AND DRAINAGE/FILTER MEDIUM

Figure 2–A–1. Liner System Designs

Source: U.S. EPA. 1988. "Draft Guidance Municipal Waste Combustion Ash," EPA/530–SW–88–006 OSWER Policy Directive No. 9573–00–1 (Washington, D.C., March 14).

REFERENCES

BNA (Bureau of National Affairs). 1987. "EPA Says Ash from Commercial Waste Burning Exempt from RCRA Hazardous Waste Regulations," *Environment Reporter* (Washington, D.C., December 4).

BNA (Bureau of National Affairs). 1988a. "EPA Gets 250 Comments on Landfill Proposal; Design, Groundwater, Financial Sections Hit," *Environmental Reporter* (Washington, D.C., December 16).

BNA (Bureau of National Affairs). 1988b. "EPA Hit for Lax Superfund Enforcement Effort at House Appropriations Subcommittee Hearing," *Environmental Reporter* (Washington, D.C., March 25).

Kovacs, William L. 1987. "Can EPA Regulate Ash as a Hazardous Waste?" *Waste Age* (Washington, D.C., May), pp. 105–111.

Lowrance, Sylvia. 1989. Statement before the Subcommittee on Transportation, Tourism, and Hazardous Materials, of the House Committee on Energy and Commerce, May 11.

Porter, Winston. 1987. Statement before the Subcommittee on Hazardous Waste and Toxic Substances, of the Senate Committee on Environment and Public Works, December 3.

Thomas, L. M. 1988. Statement before the Subcommittee on Transportation, Tourism, and Hazardous Materials, of the House Committee on Energy and Commerce, April 13.

U.S. EPA (U.S. Environmental Protection Agency). 1980. *Identification and Listing of Hazardous Waste Under RCRA, Subtitle C, Section 3001, EP Toxicity Characteristic (40 CFR 261.24)* (Washington, D.C.: Office of Solid Waste, May).

U.S. EPA (U.S. Environmental Protection Agency). 1982. *Test Methods for Evaluating Solid Waste, Physical/Chemical Methods* (SW–846), 2nd Ed., (Washington, D.C.: Office of Solid Waste, July).

U.S. EPA (U.S. Environmental Protection Agency). 1983. Letter to George W. Guhr, Chairman of Solid Waste Management Board, Waukesha County, Waukesha, Wisconsin, from John H. Skinner, Director, Office of Solid Waste (Washington, D.C., December 15).

U.S. EPA (U.S. Environmental Protection Agency). 1985. "Hazardous Waste Management System: Final Codification Rule," *Federal Register* vol. 50, no. 135, pp. 28702–28755 (Washington, D.C., July 15).

U.S. EPA (U.S. Environmental Protection Agency). 1988a. "Drinking Water Regulations: Maximum Contaminant Level Goals and National Primary Drinking Water Regulations for Lead and Copper," *Federal Register* vol. 53, no. 160, pp. 31516–31578 (Washington, D.C., August 18).

U.S. EPA (U.S. Environmental Protection Agency). 1988b. "Solid Waste Disposal Facility Criteria," *Federal Register* vol. 53, no. 168, pp. 33314–33422 (Washington, D.C., August 30).

U.S. EPA (U.S. Environmental Protection Agency). 1989. "National Primary and Secondary Drinking Water Regulations," *Federal Register* vol. 54, no. 97, pp. 22042–22160 (Washington, D.C., May 22).

3

Toxicity Testing of Ash

Because combustion destroys most organic substances, there generally has been much less concern about toxic organic materials in ash than about metals. Metals that are present in the waste stream are not destroyed through incineration and become concentrated in the ash. These can leach from ash and contaminate groundwater or surface water supplies. Two metals, lead and cadmium, have been shown to leach from ash under test conditions in quantities that cause concern and thus have been judged to have the greatest potential for harm.

The exceptions to the generalization that organic materials are not of concern are dioxins and furans (Karasek et al. 1987; Visalli 1987; Wakimoto and Tatsukawa 1985). In practice, dioxins and furans bind tightly to ash and soil particles and tend not to leach from them. Former EPA Administrator Lee Thomas summarized the agency's position on dioxins and furans in ash, saying, "Dioxins do not leach and do not appear to present a threat of ground water contamination when ash is landfilled" (Thomas 1988).[4]

There are different ways to characterize ash and assess its potential as a health risk. It can be submitted to a complete chemical analysis or subjected to tests that attempt to mimic leachate production under various field conditions. Each of these approaches, which are described below, has its strengths and limitations.

CHEMICAL ANALYSIS

Complete chemical analysis can provide information about the absolute content of metals and organics in ash. Such an analysis is helpful for assessing risk when the potential exposure to the harmful constituents in ash is direct—for instance, inhalation of ash dusts. Such an analysis provides information on the maximum amount of toxic substances that can be released to the environment, but alone it is of little value in the estimation of risk from groundwater or surface water contamination because it provides no information on the amount of hazardous constitutents likely to be liberated, that is, to leach.

Chemical analyses of incinerator ash residues have identified the presence of many substances, including the eight metals for which EPA has established regulatory (EP toxicity) limits. Inorganic chemicals found in ash in an EPA study support the generally held conclusion that concentrations of

---

[1] Because dioxins and furans generally do not leach, they do not play a prominent role in discussions about ash management, and they are not considered further in this report. However, under other circumstances—such as if particles of ash become airborne and are inhaled or injested—dioxins and furans would present risks.

most metals tend to be higher in fly ash and lower in bottom ash.[5]

Ranges for lead and cadmium are shown in table 3–1. Note that the highest value recorded for lead occurred in combined ash rather than in fly ash where concentrations are generally higher (see discussion in Appendix 3–C).

Table 3–1. Concentrations of Lead and Cadmium in Ash

	Fly Ash	Combined Ash	Bottom Ash
Lead	200-26,600	31-36,600	110-5,000
Cadmium	< 5-2,210	0.18-100	1.1-46

Note: All figures in 3 µg/g (ppm)

Source: U.S. Environmental Protection Agency (1987b). *Characterization of MWC Ashes and Leachates from MSW Landfills, Monofills, and Co–Disposal Sites*, Vol. IV, EPA 530–SW–87–028D (Washington, D.C.).

COLUMN LEACH TESTS

Column leach tests are used to model leachate production from a landfill. They involve placing a waste such as ash in a glass or plastic column, passing an extraction liquid through the column, and then analyzing the resulting leachate for various constituents. The extraction liquid can be any of a variety of solutions that are common in the environment. Tests can run for several days, several weeks, or longer. To date these tests have played no role in regulation of wastes, although some investigators claim that column tests approximate the conditions in a landfill better than the more commonly used batch tests (which are described in the next section).

Column leaching tests[6] (Francis 1984; Francis and White 1987; Francis and Maskarinec 1987;

---

[2] Appendix 3–A provides ranges of concentrations for many of the constituents that have been found in different types of MWC ash. Because of the large number of constituents of ash and the wide ranges of concentrations, these data provide only a general picture of ash composition, and it is difficult to draw conclusions from them.

[3] A more detailed description of these column leaching studies can be found in Appendix 3–B.

Cundari and Lauria 1986, 1987) analyzed leachates from combined ash that had been in contact with either a pH–neutral (distilled water) or a slightly acidic (a laboratory–derived municipal waste leachate or simulated acid rain) extraction fluid. The resulting concentrations of cadmium and lead in the leachate were all well below EP toxicity levels. As would be expected, however, the acidic extraction fluids tended to produce higher concentrations of lead and cadmium than the distilled water.

In their column leaching tests, Hjelmar (1987) and Surgi (1986) used a much higher proportion of fly ash than the 10 percent that is typically found in combined ash. In general, the more fly ash, the higher the concentration of lead and cadmium in the leachate—which supports the generalized results from the chemical composition studies of ash discussed in the previous section. When the proportion of fly ash was increased to 50 percent or 100 percent, leachate concentration surpassed the EP toxicity limit. In both studies, concentrations of lead and cadmium were highest in the initial measurements and declined as the extraction tests progressed.

Finally, Surgi (1986) conducted a column leaching test with fly ash and a highly alkaline extraction fluid (pH greater than 12.5). Because lead is amphoteric (i.e., it becomes soluble at acidic or alkaline pH, but is insoluble at neutral pH), it leaches in high quantities under these conditions. The leachability of cadmium, however, declines sharply as alkalinity rises.

BATCH LEACH TESTS

Batch tests, as with column tests, can be used to model field conditions at a landfill. They differ from the column leach tests in that the ash and the extraction fluid are vigorously mixed over a period of time in the batch tests. This maintains intimate contact between the waste and the extraction fluid for a fixed time, and the agitation of the waste solution increases the likelihood that each particle of waste is in constant contact with the extraction liquid. In general, batch tests are more rigorous than column tests. In addition, they are usually faster, easier to operate, and more reproducible than column tests (DiPietro et al. 1989). Batch leach tests play a central role in making regulatory decisions and thus most information about the leachability of metals from MWC ash comes from these tests. One type of batch test, the extraction procedure toxicity (EP tox) test, provides information about the liberation of chemicals under specified conditions and is used to judge the toxicity of wastes for regulatory purposes. Another batch test, the toxicity characteristic leaching procedure (TCLP), will soon replace the EP tox test.

EP Toxicity Test

We review the results of seven EP tox test studies carried out by government, universities, and industry (for summary, see table 3–2; for more details, see Appendix 3–C). Overall, in tests of fly ash, the cadmium levels surpassed the EP toxicity limits and in nearly all cases, lead also exceeded those limits. Levels of lead and cadmium from combined ash leachate were generally lower than concentrations from fly ash. No value for cadmium in combined ash was above the EP toxicity limits, though a few values were close. Lead values exceeded the EP toxicity limits in three of four tests. The lack of data on bottom ash make conclusions impossible; however, given the high value of lead found for one facility, additional testing is certainly warranted.

Results from these studies should not be taken to suggest that only fly ash contains concentrations of cadmium and lead that exceed the EP toxicity limits. Most notably, the highest concentration of lead in test leachate was found in the single bottom ash sample reported by EPA (1987b). Furthermore, the concentration of lead in the combined ash from Incinerator A (see table 3–2) reported by EPA exceeded the concentration of lead in fly ash from the same incinerator. One can conclude, however, that most samples, for whatever reasons, exceeded the EP tox test limits. Though the results for cadmium generally support the idea that the highest concentrations

are found in fly ash, three of four cadmium measurements of combined ash show levels that are more than half the limit values.

The large differences in lead leachate concentrations that EPA reported in duplicate samples underline the variability of test results, further complicating the drawing of conclusions from these data sets.

Toxicity Characteristic Leaching Procedure

The TCLP, another batch leach test, will eventually (possibly in 1989) replace the EP tox test as the means for determining whether a waste is toxic and thus should be classified as a hazardous waste. However, because the TCLP has not yet been adopted, there are limited data on its application to MWC ash. In three studies, both the TCLP and the EP tox test were performed on the same samples (see detailed discussion of these studies in Appendix 3–C).

In the first study (U.S. EPA 1987b), the concentrations of cadmium and lead in test leachates were similar for the two tests overall—many of the samples surpassed the regulatory limits for both lead and cadmium. However, several samples surpassed the regulatory limits in one test but not the other.

Although the EPA study did not indicate which of the two possible TCLP extraction fluids was used, two other studies indicate that the concentration of lead in leachates appears to depend heavily on which of the two extraction fluids was used in the test. Studies by Francis and Maskarinec (1987) and by Shinn (1987b) showed that the less acidic extraction fluid (#1) consistently produced lower levels of lead than the second more acidic extraction fluid (#2), which produced results similar to EP tox test results. On the basis of a pretest, the more acidic extraction fluid #2 was used on samples from the two facilities tested by Francis and Maskarinec. Thus, the resulting test leachates surpassed the regulatory limits and thus could have been classified as hazardous. In contrast, Shinn's analysis relied predominately on extraction fluid #1, and as a result, the test leachates were below the regulatory limits.

Table 3–2. Concentrations of Metals in Ash Leachates as Determined by the EP Tox Test

			Sources of Information				
	U.S. EPA (1987b)	Knudson (1986)	NYSDEC[a] (1987)	Eighmy et al. (1988)	Mika and Feder (1985)	Francis and White (1987)	Ogden Projects, Inc. (1988)
Cadmium							
Fly Ash	A 6.02 7.72 B 18.0 C 7.89 D 8.60 9.18	A (13.0) (15.0) B (69.0) C 2.25 D (100.0) E 2.0	—	—	—	—	—
Combined Ash	A 0.827 C 0.060 D 0.649	—	0.65	0.16[b]	0.67	A 0.24 B 0.5 C 0.71 D 0.02	A 0.52 B 0.47 C 0.49 D 0.41 E 0.39
Bottom Ash	B 0.388	—	—	—	0.09 0.08	—	—
Lead							
Fly Ash	A 4.72 10.8 B 19.3 C 17.8 D 19.5 25.2	A (8.0) (6.5) B (33.0) C 2.5 D (9.4) E 13.0	—	—	—	—	—
Combined Ash	A 20.8 C 2.09 D 7.25	—	6.2	3.22[b]	10.8	A 6.4 B 10.3 C 5.80 D 3.15	A 2.15 B 1.89 C 2.06 D 2.06 E 1.41
Bottom Ash	B 34.0	—	—	—	2.8 7.1	—	—

Table 3–2. (continued)

Notes: All values given in ppm (mg/l). **Bold** indicates values that exceed EP tox test limits (1.0 ppm for cadmium and 5.0 ppm for lead). Letters (A,B,C, or D) indicate different facilities where the sample was taken.

All values in parentheses indicate results from single grab samples; all other values are from analyses of composite samples.

[a] New York State Department of Environmental Conservation.

[b] This combined ash contained an unusually high proportion of bottom ash to fly ash (98–99 percent bottom ash and 1–2 percent fly ash). Typically, combined ash contains 90 percent bottom ash and 10 percent fly ash.

Sources:

U.S. EPA (U.S. Environmental Protection Agency). 1987b. *Characterization of MWC Ashes and Leachates from MSW Landfills, Monofills, and Co–Disposal Sites, Vol. V. Characterization of Municipal Waste Combustor Residues.* EPA 530–SW–87–028E (Washington, D.C.: EPA, October).

Knudson, James C. 1986. *Study of Municipal Incineration Residue and its Designation as a Dangerous Waste.* Washington State, Department of Ecology, Solid Waste Section (Olympia, Wash., August).

New York State Department of Environmental Conservation. 1987. *Ash Residue Characterization Project: Summary Report* (Albany, N.Y.: Division of Solid and Hazardous Waste, July).

Eighmy, T. Taylor, Nancy E. Kinner, and Thomas P. Ballestero. 1988. *Codisposal of Lamprey Regional Solid Waste Cooperative Incinerator Bottom Ash and Somersworth Wastewater Sludges: Final Report.* Prepared for Lamprey Regional Solid Waste Cooperative (Durham, New Hampshire, January).

Mika, S., and W. A. Feder. 1985. *RESCO Incinerator Residue Research Program: Results, Evaluations, and Recommendations: Final Report.* Prepared for Refuse–Energy Systems Company (Waltham, Mass.).

Francis, C. W., and G. H. White. 1987. "Leaching of Toxic Metals from Incinerator Ashes," *Water Pollut. Control Fed.* vol. 59, no. 11, pp. 979–986.

Ogden Projects, Inc. 1988. *Environmental Test Report No. 179,* regarding Bristol Resource Recovery Facility Residue. Prepared for Ogden Martin Systems of Bristol, Inc. (Fairfield, N.J.).

Problems with Testing Ash

For wastes not already declared hazardous, the present dichotomous system for categorizing wastes as either hazardous or nonhazardous relies predominately upon tests. Yet, for a number of reasons, efforts to characterize MWC ash with the EP tox test have produced results of questionable validity. Several studies have shown that test results are often unreproducible. For the test to be a useful regulatory tool with meaningful results, a number of criteria must be met. First, one must be able to obtain a sample that is representative of the ash waste stream. Second, test procedures must be able to characterize the ash and leachates. Finally, the quality assurance of the laboratory for conducting all aspects of the test must be sufficiently high so that results are not influenced. For all laboratory tests, meeting the first criterion is a problem. The EP tox test often fails to meet the second criterion. Little is known about the third.

Representativeness of Sample

Unlike other types of wastes, ash from MWCs is clearly *not* homogeneous. The lack of homogeneity in a facility's ash was demonstrated by Mika and Feder's (1985) study of bottom ash. If one can assume that their test procedures were internally consistent, the 96 individual samples they took over two 48–hour periods show a range of results for lead over several orders of magnitude. Clearly, obtaining a representative sample has proven to be a difficult task. The composition of municipal solid wastes generated in different geographical areas may vary, and thus the ash produced from different facilities may vary. However, there can also be variations in the ash from the same facility. Ash composition can vary over time—from season to season, day to day, and even hour to hour. The uneven distribution of some toxic constituents in the ash may be due to concentrated sources of these toxic constituents in the waste stream. For example, the largest source of lead in municipal solid waste is lead–acid automobile batteries (U.S. EPA 1989). If this source contributes heavily to the leachable lead, a single battery could considerably elevate the lead concentration in the ash near it.

Another problem with obtaining a representative sample is that ash also contains many inert—and presumably nonleachable—items such as stones and glass. Yet there is no clear procedure for how these items should be taken into account by test procedures. Presently, the EP tox test requires that large solid materials be crushed, cut, or ground to pass through a screen. Yet this procedure is impractical for some of these inert items, which can be quite large.

Some MWCs have dealt with the problem by removing and weighing the inert items and, after conducting the test, readjusting the result to reflect the proportion of inert materials. This step lowers the concentration of the toxic chemicals in the leachates. Others have simply removed and completely excluded these items from their analysis.

In summary, a valid procedure for obtaining a representative sample of MWC ash has never been developed. EPA's *Test Methods for Evaluating Solid Waste* (SW–846) describes the general procedures for developing a sampling plan yet it leaves many questions for testing ash unanswered. EPA developed a draft guidance document called "Sampling and Analysis of Municipal Refuse Incineration Ash," dated July 1987. That document addressed some of the problems, but the agency has not yet sought to finalize and release it.

EP Tox Test Discrepancies

EP tox test procedures consistently reveal troublesome variability in results, as discussed in separate studies by Shinn (1987a) and by Bleifuss et al. (1988). Both studies reported a wide range of results when splits of identical samples were analyzed by different laboratories. They conclude that laboratories interpret the EP tox test procedures differently, resulting in diverse results. Jirka et al. (1987) point to the lack of precision in the laboratory technique for adjusting pH levels. Lead, for which there was the greatest variation in test results, is very sensitive to pH levels near 5.0—the value to which the test solution must be adjusted during the extraction procedure. Yet the EP tox test allows the pH level to vary by plus or minus 0.2. As a result, slight differences in technique for adjusting the pH level—which are well within the allowable test procedures—can produce widely different results (see Appendix 3–C).

This problem with adjusting the pH level will be eliminated when the TCLP replaces the EP tox test. With the TCLP, the acid extraction fluid is added only at the beginning of the test, and no further adjustments to pH are made during the test. However, the TCLP raises a new problem for testing MWC ash. As discussed earlier, there are two possible extraction fluids that can be used in the TCLP. Shinn (1987a) has shown that when testing several samples from the same facility, different samples may not require the same extraction fluid. However, federal regulations require that several individual samples be taken and that results from these individual samples be statistically combined to determine the final test result which is then compared to the regulatory limits. Yet it makes little sense to mix individual results that were produced from different extraction fluids. At present there is no clear procedure for conducting the test under these circumstances or for interpreting the results.

Water Batch Tests

Several studies have reviewed the results of using different batch tests on MWC ash with water as the only extraction fluid. In one study, Resource Analysts, Inc. (RAI 1987) tested combined ash using a variation on the EP tox test methods; they used less acidic extraction fluids than usual, including deionized water. They found that when the pH level of the test solution was adjusted to neutral or slightly alkaline (pH 7 or 9), very little leaching of cadmium or lead occurred—the resulting leachate concentrations were all well below drinking water standards. When only water was added, however, the resulting solution was highly alkaline (pH around

12) due largely to the addition of lime to the incinerator stack for air pollution control (see table 3–C–2, Appendix 3–C). Cadmium concentrations fell further as the pH level rose. However, in agreement with Surgi's (1986) column test results using a very alkaline extraction fluid, the leaching of lead increased when the pH was near 12.

In a separate study, the U.S. EPA (1987b) (see Appendix 3–C) used a water batch test called the solid waste leaching procedure (also called SW–924; see U.S. EPA 1984). There are three differences between this test and the EP tox test: distilled water is used as the extraction fluid instead of an acid; only 10 times the waste's weight is added as opposed to 20 times its weight; and the extraction step can be repeated several times using the same waste sample instead of just once. EPA performed the extraction step twice on samples from four different facilities (see table 3–C–4). In all cases, they found that leachate concentrations were well below EP toxicity levels, and in most cases they were also below drinking water standards. Although pH values were not reported from these extractions, none of these facilities used scrubbers. In general, the use of scrubbers at MWCs produces very alkaline ash and so it is not likely that pH values in the EPA study ever approached those in the RAI study. This difference would explain the low concentrations of lead in the EPA study leachates.

FIELD DATA

Measurements of actual field leachates collected from disposal facilities provide the most direct information on the quantity of toxic constituents leaching from ash. Unfortunately, few data of acceptable quality are available. Information that characterizes the ash going into the landfill and that enumerates the other wastes, if any, that have been codisposed with the ash often is lacking. Furthermore, no data have been collected over periods longer than 10 years. As a result, it is difficult to use field leachates to draw conclusions about the likelihood of future contamination problems.

Field samples from three ash monofills (Hjelmar 1987; Malcolm Pirnie, Inc. 1988; Sears 1988) showed concentrations of metals in the leachate to be well below current regulatory levels for the EP tox test. However, several samples for lead (Malcolm Pirnie, Inc. 1988; Sears 1986) were above or near 0.5 ppm (an order of magnitude below the current standard). A fourth study by the U.S. EPA (1987b) showed measurements of five metals to be well below EP toxicity levels. But two measurements for lead (out of nine total samples) taken from two different facilities were 1.33 ppm and 2.92 ppm—below the EP toxicity level (5.0 ppm for lead) but considerably higher than results from the other field studies. However, EPA is not certain that these samples represent actual field leachate; instead, they may be surface runoffs.

EPA also sampled leachates from two codisposal landfills (U.S. EPA 1987c). These leachates contained metals near or below drinking water standards. However, data analyzed by Hjelmar (1987) and Denison (1988) showed that codisposal of municipal solid waste and/or municipal solid waste leachates with ash in a landfill resulted in concentrations of metals in the leachate that were higher than when the waste or municipal solid waste leachates were not present. The higher levels could be due to either greater leaching from ash during codisposal or from the municipal solid waste alone.

Only Hjelmar's study presents data over an extended period—but just 10 years. His results show concentrations of certain constituents, including lead, rising over the last 2 years of the sampling period, but it is difficult to draw conclusions from these data because only one sample was taken each year. Hjelmar's field findings also contradict column tests which show that ash tends to leach the most upon initial contact with the extraction fluid and then leach very little thereafter. Also, Cundari-and Lauria's (1986) column test which simulated 25 years of leaching showed that concentrations of lead and cadmium remained fairly constant over the test period but both declined slightly during the last 5 "years" of the simulation. A more detailed discussion of field studies can be found in Appendix 3–D.

SUMMARY

A number of studies have been performed to better define the conditions under which lead and cadmium leach from MWC ash. There is wide agreement that acidic conditions cause far greater leaching of lead and cadmium than pH–neutral conditions.

Even though metals tend to be concentrated in the fly ash, some samples of all types of ash (fly, combined, and bottom) have been found to exceed the EP tox test limits, predominately because of elevated levels of lead and cadmium. Although the EP tox test is an important regulatory tool, it has been shown to characterize ash inaccurately, and results depend heavily on how the analysis was conducted.

Several laboratory studies have shown that under very alkaline conditions, ash will leach significant quantities of lead but not cadmium. Because some ash is highly alkaline, when it comes in contact with water it may produce leachate with elevated (i.e., above EP toxicity) levels of lead. High alkalinity results from lime used in scrubbers that ends up in fly ash. Although few existing incinerators have scrubbers (U.S. EPA 1987d), this is changing in anticipation of stricter federal standards for airborne emissions.

Actual field data are limited. In the short term, most field tests have shown that very little lead or cadmium has leached from ash monofills. In fact, many field data concentrations are 100–fold less than the EP toxicity limits and below drinking water standards. Little, however, is known about how monofilled ash is likely to behave over the long term.

Parameter	Fly Ash	Combined Bottom and Fly Ash	Bottom Ash
Arsenic	15-750	2.9-50	1.3-24.6
Barium	88-9,000	79-2,700	47-2,000
Cadmium	<5-2,210	0.18-100	1.1-46
Chromium	21-1,900	12-1,500	13-520
Lead	200-26,600	31-36,600	110-5,000
Mercury	0.9-35	0.05-17.5	ND-1.9
Selenium	0.48-15.6	0.10-50	ND-2.5
Silver	ND-700	0.05-93.4	ND-38
Aluminum	5,300-176,000	5,000-60,000	5,400-53,400
Antimony	139-760	<120-<260	
Beryllium	ND-<4	ND.1-2.4	ND-<0.44
Bismuth	36-<100		ND
Boron	35-5,654	24-174	85
Bromine	21-250		
Calcium	13,960-270,000	4,100-85,000	5,900-69,500
Cesium	2,100-12,000		
Cobalt	2.3-1,670	1.7-91	3-62
Copper	187-2,380	40-5,900	80-10,700
Iron	900-87,000	690-133,500	1,000-133,500
Lithium	7.9-34	6.9-37	7-19
Magnesium	2,150-21,000	700-16,000	880-10,100
Manganese	171-8,500	14-3,130	50-3,100
Molybdenum	9.2-700	2.4-290	29
Nickel	9.9-1,966	13-12,910	9-226
Phosphorus	2,900-9,300	290-5,000	3,400-17,800
Potassium	11,000-65,800	290-12,000	920-13,133
Silicon	1,783-266,000		1,333-188,300
Sodium	9,780-49,500	1,100-33,300	1,800-33,300
Strontium	98-1,100	12-640	81-240
Tin	300-12,500	13-380	40-800
Titanium	<50-42,000	1,000-28,000	3,067-11,400
Vanadium	22-166	13-150	53
Yttrium	2-380	0.55-8.3	
Zinc	2,800-152,000	92-46,000	200-12,400
Gold	0.16-100		
Chloride	1,160-11,200		
Country	USA, Canada	USA	USA, Canada

ND - Not detected at the detection limit
Blank - Not reported, not analyzed for

Appendix 3–A. Range of Concentrations of Inorganic Constituents in MWC Ash

Note: All values given in µg/g (ppm). ND – Not detected at the detection limit.
Blank – Not reported, not analyzed for.

Source: U.S. EPA, *Characterization of MWC Ashes and Leachates from MSW Landfills, Monofills, and Co–Disposal Sites*, Summary, Vol. 1, EPA/530–SW–87–028A (Washington, D.C.: EPA, October).

APPENDIX 3–B

COLUMN LEACH TESTS

Francis (1984) and Francis and White (1984)

Francis and White designed a column test to simulate leaching of ash under field conditions. Composite samples of combined ash from four incinerators were placed in columns of 38.7 centimeter diameter and 30.5 centimeter height; each column contained between 3.3 and 3.6 kilograms of ash. Francis and White used two types of extraction liquids to mimic conditions at codisposal and monofill disposal sites: a laboratory–derived municipal waste leachate and distilled water. The liquid–to–solid ratio was about 28 to 1 (by weight), and the total volume of liquid was passed through the column over a period of about 15 weeks. Leachate samples were collected twice weekly the first month and weekly thereafter, and the samples were combined at five intervals over the test period. Francis (1984) analyzed the leachate for five of the eight regulated metals. As might be expected, levels of metals from ash leached with water were considerably lower than those leached with the laboratory–derived municipal waste leachate. Levels of the five regulated metals in the leachate were consistently below drinking water standards except for one measurement of selenium at 11 ppb, which was 1 ppb above the 10 ppb drinking water standard. Most measurements of the metals concentration were at or below drinking water standards when ash was extracted with municipal waste leachate. A few individual measurements were above the drinking water standard, but all were at least one order of magnitude below EP toxicity limits (see table 3–B–1 for lead and cadmium results.)

Table 3–B–1. Results of Combined Ash Column Leach Tests

	Francis and White (1987)	Francis and White (1987)	Francis and Maskarinec (1987)	Cundari and Lauria (1986)	EP Toxicity Limit
Extraction liquid	distilled water	municipal waste leachate	municipal waste leachate	simulated acid rain	
CADMIUM					
highest	0.0035	0.0555	0.11	0.05	1.0
lowest	<0.0005	0.0002	0.023	0.01	
LEAD					
highest	0.005	0.0570	<1.20	0.13	5.0
lowest	<0.001	0.0035	<0.40	<0.05	

Note: All values given in ppm (mg/l).

Sources: Francis, C. W. and G. H. White. 1987. "Leaching of Toxic Metals from Incinerator Ashes," *J. Water Pollut. Control Fed.*, vol. 59, no. 11, pp. 979–986.

Francis, C. W. and M. P. Maskarinec. 1987. *Leaching of Metals from Alkaline Wastes by Municipal Waste Leachate*, Environmental Sciences Division Publ. No. 2846. Prepared for the U.S. Environmental Protection Agency (Washington, D.C., March).

Cundari, Kenneth L., and Jeffrey M. Lauria. 1986. "The Laboratory Evaluation of Expected Leachate Quality from a Resource Recovery Ashfill." Presented at the Triangle Conference on Environmental Technology (Chapel Hill, N.C.).

Francis and Maskarinec (1987)

In a similar experiment, Francis and Maskarinec tested combined ash from two of the four incinerators used in Francis and White's experiment. Both experiments used a similar laboratory–derived municipal waste leachate. Francis and Maskarinec's experiment differed, however, in that it used an upflow column rather than top–to–bottom flow; leachate was pumped into the bottom of that column and then drawn up and out the top. An upflow column increases contact time between ash and the extraction liquid and may explain in part the higher concentrations of cadmium that were found in comparison to those reported by Francis and White. Francis and Maskarinec analyzed the leachate for 27 elements (see table 3–B–1 for the range of values for lead and cadmium).

Cundari and Lauria (1986, 1987)

In yet another column leaching experiment, Cundari and Lauria developed a simulation of leaching over 25 years at a monofill containing combined ash. Several columns were placed in sequence, and composite samples of combined ash representing 15 years' worth of ash disposal were placed in the columns. Next, a synthetic acid rain (pH 4.2, below the pH 5.0 used in the EP tox test)

was used to leach the ash; the liquid–to–ash–ratio was based on the precipitation rate (about 50 inches per year) and the depth of the ash at the actual monofill location. Finally, additional ash was added incrementally to the columns to simulate yearly additions of ash to the monofill. Cundari and Lauria took leachate samples after the equivalent of a year's precipitation had passed through the column. The resulting analysis of the leachate for the eight regulated metals showed levels of at least an order of magnitude below the EP toxicity limits, but considerably higher than levels recorded by Francis and White. The higher levels probably reflect the fact that the acidic "simulated acid rain" is a more effective extraction liquid than distilled water. The highest value was 0.05 ppm for cadmium and 0.13 ppm for lead (table 3–B–1). These peak values occurred in the years designated as comparable to the third and the ninth under field conditions. Although 3 to 5 times greater than the drinking water limits, these concentrations were still 20 to 38 times less than the EP toxicity limits.

Hjelmar (1987)

Hjelmar conducted column leaching tests using 10 combinations of fly ash and bottom ash in varying proportions (see table 3–B–2). Ash samples

Table 3–B–2. Column Leach Test Results with Different Ash Compositions

Metals	Facility	Leachate Concentration			EP Toxicity Limits
		100% fly ash	49–51% fly ash 51–49% bottom ash	13–15% fly ash 87–85% bottom ash	
CADMIUM					
	A	0.019	0.51	0.0054	1.0
	B	7.1	1.4	0.012	
LEAD					
	A	<0.1	0.05	<0.005	5.0
	B	0.41	1.3	<0.03	

Notes: Values given in μg/kg ash.

Source: Hjelmar, Ole. 1987. "Leachate from Incinerator Ash Disposal Sites." Presented at International Workshop on Municipal Waste Incineration (Montreal, Canada, October 1–2).

consisted of composites collected over a 1–week period from two facilities. Hjelmar used an upflow column with either artifical rainwater or seawater as the extraction liquid. Two general observations can be drawn from his results: the leachate concentrations of six metals (arsenic, cadmium, molybdenum, lead, selenium, and zinc) usually increased as the proportion of fly ash increased; and leachate concentrations were highest when the liquid–to–solid ratios were very low (0.1 or less).

Hjelmar also presented data that compared the concentration of metals in the leachate to the liquid–to–solid ratio. These data showed that the concentration of metals almost always peaked at the initial measurement—when the liquid–to–solid ratio was very low.

Surgi (1986)

Surgi reported results from leaching fly ash in a column with synthetic acid rain at pH 3.93. The liquid–to–solid ratio was low (about 2), and leaching occurred over approximately an 8–hour period, with samples collected about every 40 minutes. The results showed a large range in metal concentrations. Cadmium ranged from an initial measurement of greater than 10 ppm downward to below the drinking water standard (0.01 ppm). Lead ranged from slightly less than 1 ppm to below 0.05 ppm (the drinking water standard). In agreement with Hjelmar's findings, the highest values were found in the initial measurements. After the initial peak value, the concentration of both cadmium and lead in leachate declined rapidly—both were below 0.05 ppm within 3 hours. The pH remained constant at approximately 10 during the entire 8–hour experiment, which is explained by the high pH of the ash.

Surgi conducted another experiment to observe the effect of an alkaline solution on the fly ash. Surgi added excess base (calcium oxide) to maintain the pH above 12.5 in the extraction liquid. Results of this experiment showed that the leachability of cadmium declined dramatically—the concentration of cadmium in the leachate remained consistently near or below the drinking water limits. As

expected, the concentration of lead (which is soluble at acid as well as alkaline pH) in the leachate was very high—well above the EP toxicity limits, and initial measurements were over 10 times higher than acceptable limits.

APPENDIX 3–C

BATCH LEACH TESTS

EP TOX TEST RESULTS

Data from using the EP tox test to extract metals from ash residues thus far show that the only elements that sometimes surpass EP toxicity limits are cadmium and lead (limits 1.0 ppm and 5.0 ppm, respectively) (see table 3–2). For the other six metals, most measurements are at or below drinking water standards.

U.S. EPA (1987b)

Versar, Inc., a consulting firm, sampled three types of ash from four incinerators under an EPA contract. They collected samples of fly ash and either a bottom ash or combined ash sample from each incinerator. Large items such as cans and metal scraps were excluded from the samples. Combined ash was sampled at three of the four facilities because it was impossible to collect a separate bottom ash sample. Next, to produce time–composite samples, they combined periodic grab samples—typically, hourly samples over an 8–hour shift. By varying the day, shift, and/or incinerator unit from which samples were taken, four time–composite samples were obtained from each incinerator. A single final sample was derived from these time–composite samples. Duplicate final samples, which were analyzed for two of the four fly ash samples, were averaged. The results show that all fly ash values for cadmium exceeded EP toxicity limits, including one value that was 18 times the standard. The combined ash and bottom ash samples were below the EP toxicity limits. For lead, all four values for fly ash were above EP toxicity limits, as well as two of three values for combined ash. The one lead measurement for bottom ash, 34.0 ppm, was also well above the EP toxicity limits. Interestingly, that measurement was higher than any from combined or fly ash. The high concentration might have resulted from remains of lead–containing items such as batteries being picked up in the bottom ash sample, but this Versar report advances no explanation (U.S. EPA 1987b).

Knudson (1986)

Knudson reported test results for fly ash from several incinerators for the state of Washington. The results represent five different incinerators. Composite samples were tested from two incinerators for lead and cadmium, and single-grab samples were tested from the other three. The composite samples were comprised from either eight grab samples taken every half hour over a 4–hour period or from hourly grab samples over a 24–hour period. For cadmium the two values taken from composites were above the EP toxicity limit; however, the grab samples were all considerably higher, including one measurement that exceeded the limit by a factor of 100. For lead, one of the composite values surpassed the EP toxicity limit and one did not. All grab samples were above the limit for lead.

New York State Department of Environmental Conservation (1987)

The New York State Department of Environmental Conservation sampled combined ash at the Oswego County incinerator facility. It first took grab samples from every truckload of ash (6 to 7 truckloads per day) generated at the facility for 7 consecutive days. Large, nonfriable material was removed from the sample. Daily composites consisted of mixtures of each day's grab samples, and the investigators prepared five replicate samples of each daily composite. Daily results, the average values of the five composites, ranged from 0.41 to

1.15 ppm for cadmium and from 3.1 to 9.6 ppm for lead. The final results, 0.65 ppm and 6.2 ppm for cadmium and lead, respectively, are the average of the seven daily averages.

Eighmy, et al. (1988)

Eighmy et al. sampled bottom ash from a resource recovery facility in Durham, New Hampshire. Bottom ash at this facility is estimated to comprise 98–99 percent of the total ash residue, with fly ash comprising only the remaining 1–2 percent. To form a composite, they took five samples of bottom ash from an ash pile that represented about 6 months of ash production. Next, they took random subsamples of the composite and analyzed 9 for cadmium and 17 for lead with the EP tox test. The results from the subsamples were averaged to obtain the final values of 0.16 ppm for cadmium and 3.22 ppm for lead, both of which are below EP toxicity limits. Concentrations of cadmium in the nine subsamples tested for cadmium ranged from 0.13 to 0.20 ppm, all below the EP toxicity limits. One–fourth of the lead concentrations from the subsamples, which ranged from 0.76 to 13.0 ppm, were above the EP toxicity limits. They note that one group of samples had a higher mean which could have been due to the differences in test procedures.

Mika and Feder (1985)

Mika and Feder sampled bottom ash and combined ash at the RESCO facility in Saugus, Massachusetts. The bottom ash samples were collected hourly over two 48–hour periods in 1982. Each hourly sample was analyzed and each set representing samples over a 48–hour period was averaged to produce two values. For cadmium, both averaged values—0.09 ppm and 0.08 ppm—were well below EP toxicity levels. However, the 96 samples ranged from below detection limits to 1.59 ppm, though the second highest value was < 0.6 ppm. For lead, one averaged result (7.1 ppm) was above the regulatory level and one value (2.8 ppm) was below. The individual samples ranged from 0.04 to 33 ppm.

For the combined ash, duplicate hourly composite samples were taken on 12 days throughout a 6–week period, and three different laboratories participated in the analysis of the samples. Two laboratories analyzed each of the samples in duplicate, producing eight values for each of the original 12 samples. The final values that appear in table 3–2 are the averages of analyses of nearly 100 individual samples. For cadmium, the average was below EP toxicity limits at 0.67 ppm. For lead it was over twice the limit at 10.7 ppm.

Francis and White (1987)

Francis and White obtained composite samples of combined ash from four incinerator facilities in four different states. (The authors do not explain their composite sampling procedures.) Extraneous materials, which represented between 3 and 36 percent of the samples by weight, were removed and the remaining ash was milled to < 2 millimeters (considerably smaller than the 9.5 millimeters required for the standard EP tox test). The milling process, presumably, increased the leachability of constituents. None of the samples exceeded the EP toxicity levels for cadmium. However, three of four samples were above the regulatory level for lead.

Ogden Projects, Inc. (1988)

Versar, Inc., sampled combined ash from the Ogden Martin Resource Recovery Facility in Bristol, Connecticut, in a study conducted for Ogden Projects, Inc. Over a period of 4 days in April 1988, they took 26 samples. In addition, they made four composite samples, two from the field and two from the laboratory. For field composites, eight successive individual samples of ash were combined in a bucket; for laboratory composites, 100–gram aliquots of dried and crushed ash from the same individual samples were combined. Inert materials such as glass were sifted out of the sample and carefully weighed. Also, samples were dried to remove

excess moisture, and their initial weight was noted. After the extraction was completed, the results were adjusted to allow for the excluded inert material and excess moisture. None of the samples exceeded the EP toxicity limits.

TOXICITY CHARACTERISTIC LEACHING PROCEDURE

We reviewed three studies that used both the TCLP and the EP tox test to evaluate MWC ash. Test results for lead and cadmium from the first two of these studies are summarized in table 3–C–1 (U.S. EPA 1987b). In EPA's study, samples from four facilities were evaluated (see EP TOX TEST RESULTS section at the beginning of Appendix 3–C for a description of the sampling methods and test results). EPA found that concentrations of toxic chemicals in the TCLP leachates were similar to the EP tox test results in that many, but not all, of the samples surpassed the regulatory limits. Beyond this initial observation, it is difficult to draw clear conclusions. This was the only study of three that tested fly ash, combined ash, and bottom ash separately. The concentrations of cadmium in fly ash leachates tended to be much higher than in the combined or bottom ash, but one cannot draw this conclusion from the results for lead. At one facility, for example, leachates from the fly ash were below regulatory limits for lead, but well above for combined ash. When these TCLP results are compared with those from the EP tox test (see table 3–2), there is some correlation between results for the two tests, but not in all cases.

Unfortunately, EPA's study does not indicate which of the two TCLP extraction fluids was used in this test. As discussed in Chapter 2, extraction fluid #1 contains the same amount of acid added in the EP tox test; extraction fluid #2 contains much more acid and is used with certain highly alkaline wastes. EPA included the second, more acidic extraction fluid because continuous contact with an acidic leachate may eventually exhaust the alkalinity of even highly alkaline wastes—such as some MWC ash. One determines which extraction fluid to use by conducting a pretest on a small waste sample.

Francis and Maskarinec (1987)

In this study, Francis and Maskarinec tested a composite ash sample from two incinerators (see table 3–C–1). Although the pretest of these samples showed that the second (more acidic) extraction fluid should be used, separate tests were run with both extraction fluids on the same ash samples. Furthermore, Francis and Maskarinec conducted the EP tox test on the samples. For cadmium, results from the TCLP and the EP tox test where all below regulatory limits. In results from one incinerator, the less acidic extraction fluid (#1) produced leachates much lower (by an order of magnitude) than the EP tox test or the TCLP with extraction fluid #2, but the results from the other incinerator for all three tests varied very little.

Table 3–C–1. Concentrations of Cadmium and Lead in Ash Leachates as Determined by the TCLP

	U.S. EPA (1987b)				Francis and Maskarinec (1987)			
	TCLP		EP Tox Test		TCLP			EP Tox Test
					Extraction Fluid			
					#1	#2		
CADMIUM								
Fly Ash	A 0.015	0.032	6.02	7.72				
	B 17.2		18.0					
	C 8.36		7.89					
	D 10.3	8.90	8.60	9.18				
Combined ash	A 0.682		0.827		A 0.03	0.31		0.179
	C 3.32		0.060					
	D 0.025		0.649		B 0.31	0.73		0.425
Bottom ash	B 0.418		0.388					
LEAD								
Fly Ash	A 0.962	<0.05	4.72	10.8				
	B 6.91		19.3					
	C 13.4		17.8					
	D 15.2	12.8	19.5	25.2				
Combined ash	A 16.3		20.8		A <1.5	6.6		5.83
	C 7.30		2.09					
	D 0.655		7.25		B 1.25	7.2		12.0
Bottom ash	B 30.1		34.0					

Notes: Values given in ppm (mg/l); letters (A,B,C,D) indicate facilities from which samples were taken.

Sources: U.S. Environmental Protection Agency. 1987. *Characterization of MWC Ashes and Leachates from MSW Landfills, Monofills, and Co–Disposal Sites, Vol. V. Characterization of Municipal Waste Combustor Residues.* EPA 530–SW–87–028E (Washington, D.C., EPA, October).

Francis, C. W., and M. P. Maskarinec. 1987. *Leaching of Metals from Alkaline Wastes by Municipal Waste Leachate*, Environmental Sciences Division Publ. No. 2846. Prepared for the U.S. Environmental Protection Agency (Washington, D.C., March).

For lead, the results from the EP tox test with the more acidic extraction fluid (#2) were similar and surpassed regulatory limits, whereas results from the TCLP using the less acidic extraction fluid (#1) were below regulatory limits. Since the pretest indicated that the more acidic extraction fluid would have been required, this ash could have been classified as a hazardous waste using either test.

Shinn (1987b)

In a third study, Shinn analyzed the same 18 samples with the TCLP and the EP tox test, on behalf of the Oregon Department of Environmental Quality (DEQ). On the basis of a pretest, Shinn found that most of the samples required extraction fluid #1; however, three samples required extraction fluid #2. Shinn reports that separate tests using both extraction fluids on the three samples with high pHs were conducted. Only extraction fluid #1 was used on the remaining 15 samples.

As required by EPA test procedures in SW–846 (1982), Shinn then took the upper confidence limit of the leachate concentration from the 18 samples and compared this value to the regulatory limit. When extraction fluid #1 was used on all 18 samples, the results for the concentration of lead in the leachates were < 0.48 ppm and < 0.98 ppm, with the first result reported by the DEQ and the second by a private laboratory. When the three samples that required extraction fluid #2 were included, the upper confidence limit rose to < 1.25 ppm (analysis by DEQ only).

EP TOX TEST DISCREPANCIES

Bleifuss et al. (1988)

Even if samples are identical, variations in test procedures can produce different results. A study by Bleifuss et al. demonstrated some of the problems with reproducing EP tox test results, particu-

larly for lead. The study also reported a wide range of results from tests conducted at several different laboratories. Fly ash was chosen because it is generally more homogenous than other types of ash and therefore tends to reduce sampling error. They prepared three samples using different proportions of fly ash from three facilities. The resulting three composites were then individually mixed and splits of the composites were sent to 13 different laboratories. Using the EP tox test, the laboratories analyzed the splits for five metals. Several laboratories performed the test two or three times on their samples. Test results for lead varied widely among the different laboratories (see table 3–C–2). The variation for lead was less for individual labs conducting two or three tests but much greater than the variation in results for other metals. The authors conclude that the EP tox test is unreliable for lead and thus single test results are meaningless.

Ogden Projects Inc. (1986)

In a similar study, Ogden Projects Inc. performed tests on combined ash from a waste–to–energy incinerator in Brooks, Oregon. First, three 8–hour composites were taken; after mixing, each of the three composites was divided into an imaginary matrix or sampling grid. Next, six samples were taken from random locations in the grid to produce a total of 18 samples. In the first round of analysis, Brown & Caldwell, a private testing laboratory, and the state DEQ laboratory, separately tested splits of the same eight samples (a subset of the 18 samples) with the EP tox test. A third laboratory, Coffey, a private testing laboratory, analyzed all 18 samples. The laboratories tested the ash for the eight EP tox test metals; only lead surpassed the regulatory standard. Two months later, they conducted a second round of analysis using additional sample sets derived from the original 18 samples. This time all three labs analyzed the entire 18 samples (see table 3–C–3 for results for lead).

Table 3–C–2. Results of EP Tox Test of Fly Ash for Lead from 13 Different Laboratories

Lab No.	Fly Ash Composite No. 1			Fly Ash Composite No. 2			Fly Ash Composite No. 3		
1	26			14			33		
2	22			13			29		
3	12.3	10.6	14.0	12.6	13.9	18.7	10.6	13.7	11.6
4	26			19			31		
5	10.0	11.0	15.0	28.0	18.0	16.0	18.0	20.0	16.0
6	7.7			11			2.3		
7	26			14			28		
8	2.19			15.3			2.06		
9	33	37		30	34		49	41	
10	5.1	3.9	2.8	2.7	2.8	2.4	2.0	2.4	2.9
11	20	24	24						
12	24.8	29.8	27.7	19.9	22.0	22.0	25.3	13.9	10.6
13	28.0			43.2			8.0		
$\bar{x}$	19.34			18.43			18.90		
σ	10.45			10.47			14.1		
Max	37			43.2			49		
Min	2.19			2.4			2.0		

Note: Values given in ppm (mg/l)

Source: Bleifuss, Rodney L., Harlan B. Niles, and John Engesser 1988. "Round Robin Analyses (1) EP Toxicity Test (2) ASTM Water Leach Test," Technical Report NRRI/GMIN–TR–88–06, prepared for the Minnesota Waste to Energy Association by the Natural Resources Research Institute, University of Minnesota, Duluth, Minnesota., January 15.

The wide range of results within each laboratory's sample set implies a lack of homogeneity in the ash sampling grid. The variation among the three laboratories analyzing splits of the same sample also could be due in part to a lack of homogeneity. But more likely, as Shinn of the DEQ staff concludes, "there appears to be a distinct bias between [sic] laboratories" (Shinn 1987a, p. 15), which could be attributed to differences in execution or interpretations of the EP tox test. Shinn further points out that certain aspects of ash not addressed by the EP tox test protocol are likely sources of some of this test result variability. These are (1) the inclusion of heavy inert material in the sample which would displace some of the leachable ash and thus lower the concentration of ash leachate; (2) the inclusion of waste water (such as quench water) in the sample which would also displace some of the leachable ash and could lower the result; and (3) the upper limit on the extraction acid, which could result in a pH above 5.0 for very alkaline waste. From this study, Shinn concludes that "further sampling and analysis will not improve the precision and accuracy... However it is impossible to state with any confidence whether or not the waste exceeds the regulatory threshold [i.e., the EP tox limits.]" (Shinn 1987a, p. 17).

Shinn notes in a follow–up to this study (1987b) that the method of pH adjustment can also cause variability in test results. This finding is echoed by a study conducted by the EPA's Central Regional Laboratory in Chicago (Jirka et al. 1987) which sought to explain the wide variability in EP tox test results for metals. EPA, as did the Oregon DEQ, observed inter– and intralaboratory variability analyses of duplicate samples, which were too great to be explained by normal sampling error. After conducting several experiments they found that the large variability in EP tox test results could be due to the technique for adjusting pH. When small aliquots of acid were added with constant

Table 3–C–3. EP Tox Test Results for Lead in Ash from Marion County Resource Recovery Facility

Sample	Coffey Nov. 1986	Coffey Jan. 1987	Oregon DEQ[a] Nov. 1986	Oregon DEQ[a] Jan. 1987	Brown Caldwell Nov. 1986	Brown Caldwell Jan. 1987
1–1	4.10	2.40	21.0	4.0		4.0
1–2	3.13	0.864	7.8	5.2	1.5	5.7
1–3	0.56	0.193		1.4		1.1
1–4	2.19	0.480		2.1		3.5
1–5	2.34	1.75	1.8	5.4	3.8	2.3
1–6	1.14	0.234		1.0		0.48
2–7	3.39	2.17		2.8		2.2
2–8	2.12	1.56		3.2		1.8
2–9	5.16	4.50	11.0	5.5	4.2	3.0
2–10	4.34	1.54		3.4		1.9
2–11	3.42		0.956	5.2		1.8
2–12	21.00	17.1	93.0	26.0	2.3	16.0
3–13	1.05	1.01	9.0	3.4	19.0	1.3
3–14	1.64	0.634		4.8		1.4
3–15	0.74	0.580		8.6		1.4
3–16	3.26	0.550	4.1		1.3	4.5
3–17	1.07	0.532	7.2	1.5	24.0	3.5
3–18	2.47	1.50	3.5	3.7	2.3	4.1

Note: Coffey, Oregon DEQ, and Brown Caldwell are the three laboratories that conducted the analysis. Values given in ppm (mg/l).

[a] Oregon Department of Environmental Quality

Source: Shinn, Claude E. 1987. *Toxicity Characteristic Leaching Procedure (TCLP), Extraction Procedure Toxicity (EPtox), and Deionized Water Leaching Characteristics of Lead from Municipal Waste Incinerator Ash*. Oregon Department of Environmental Quality, Laboratories and Applied Research Division (Portland, Oregon, July), Appendix C.

mixing, the amount of metal extracted was much lower than when larger aliquots were added without constant mixing. In the latter case, Jirka et al. postulate that pH values in some areas of the sample actually fall below the 5.0 plus or minus 0.2 pH range required by the test. Because the solubility of lead is very sensitive to pH, the adjustment technique could greatly influence the results. The solubility of lead was found to be particularly sensitive to pH near 4.8, the lower limit in the EP tox test. The authors conclude with a recommendation that the EPA specify the technique for pH adjustment during the test.

WATER BATCH TESTS

A series of EP tox tests at the Claremont, New Hampshire, facility provides considerable information about how pH levels affect the amount of metals extracted. These tests were conducted by Resource Analysts, Inc. (RAI 1987) for Signal Environmental Systems, in accordance with the conditions by which the landfill received its permit to accept ash. RAI collected grab samples every 10 minutes for a 1–hour interval on 3 different days during 1987. They analyzed the three composite daily samples using EP tox test procedures, except that the pH was adjusted to three levels—9, 7, and 5 (see table 3–C–4). Several tests were also conducted where the EP tox test procedures were followed except that no acid was added to the extraction fluid. The pH of the samples mixed only with water ranged from about 11.9 to 12.7. The high alkalinity of these ash samples was due in part to the lime added to the air pollution control devices.

Results showed that the level of cadmium extracted from the ash increased as the pH declined.

Only at pH near 5 did the cadmium concentration approach the EP toxicity limit. In 2 samples out of 42 (including duplicate samples), the concentration surpassed the regulatory limit. For lead, the levels were lowest at pH 9 and 7. At pH 5 and 12, the amount of lead extracted rose dramatically—at both pH levels; leachate concentrations averaged above EP toxicity limits.

Table 3–C–4. EP Tox Test Results with Different Extraction Solution pH

	pH				EP Toxicity Limits
	5	7	9	12	
CADMIUM					
Average	0.45	0.14	0.044	< 0.005	1.0
Lowest	0.081	0.018	< 0.005	< 0.005	
Highest	1.2	0.36	0.32	< 0.005	
LEAD					
Average	7.6	0.22	0.077	8.1	5.0
Lowest	1.2	< 0.05	< 0.05	5.2	
Highest	32	0.37	0.24	15	

Note: All values in ppm (mg/l).

Source: Resource Analysts, Inc. 1987. Unpublished data from Claremont, New Hampshire, SES Resource Recovery Facility.

Table 3–C–5. Concentrations of Cadmium and Lead from the Solid Waste Leaching Procedure

Facility	Ash Type	Cadmium		Lead	
		No. 1	No. 2	No. 1	No. 2
A	Fly	<0.015	<0.01	<0.075	0.072
	Fly (duplicate)	<0.01	<0.01	<0.05	<0.05
	Combined	<0.01	<0.01	<0.05	<0.05
B	Fly	<0.01	0.033	<0.05	0.148
	Bottom	<0.01	<0.01	<0.05	<0.05
C	Fly	0.122	<0.01	0.128	<0.05
	Combined	<0.01	<0.01	0.063	<0.05
D	Fly	<0.015	<0.01	<0.05	<0.05
	Fly (duplicate)	<0.01	<0.01		
	Combined	<0.01	<0.01	<0.05	<0.05

Note: Two extractions, conducted in sequence, were taken from the same sample and are shown in columns No. 1 and No. 2. Values in ppm (mg/l).

Source: U.S. Environmental Protection Agency 1987, *Characterization of MWC Ashes and Leachates from MSW Landfills, Monofills and Co–Disposal Sites*, Vol. V, *Characterization of Municipal Waste Combustor Residues*. EPA/530–SW–87–028E (Washington, D.C., EPA, October).

EPA developed the solid waste leaching procedure (SWLP, also called SW–924; see U.S. EPA 1984) to evaluate a facility where no municipal solid waste would be disposed of, such as at a monofill. Although this procedure exists in draft form, it has never been published as a final document. The test is conducted by mixing a waste sample with 10 times its weight of pure water, mixing the solution for 18 hours, and then removing and analyzing the leachate. Next, more water is added to the same waste and the test is repeated. SW–924 suggests four sequential leachings, but only two were performed in this study. A discussion of how the ash samples were collected can be found in the description of EP tox test results (see U.S. EPA 1987b) at the beginning of this appendix. As with the EP tox test and the TCLP results, four facilities were sampled and where possible, different types of ash were sampled. Results are presented in table 3–C–5. Concentrations of SWLP leachates were all below EP toxicity limits, and most were also below drinking water standards. The highest values for cadmium (from fly ash) and lead (from combined ash) were about 10 times below EP toxicity limits.

APPENDIX 3–D

FIELD DATA

Measurements of metal concentrations in leachates from waste sites can provide information about what actually happens in the environment and also provide a basis for comparison with predictions made from results of the EP tox test or other tests. Useful as such measurements can be, only a limited number are available from MWC ash landfills.

Hjelmar (1987)

Hjelmar reported field data from two ash disposal sites in Denmark on behalf of the Danish Water Quality Institute (DWQI). The first site, a mono-fill, contained only combined ash, which Hjelmar estimates to contain 10 to 40 percent fly ash. The DWQI sampled leachate at this facility from 1973 to 1986, but data from 1977–1980 were deemed unreliable by Hjelmar and were excluded. The DWQI collected leachate from a drainage system above the liner and analyzed the leachate for several constituents including five EPA–regulated metals. None of these metals was present in concentrations that approached EP toxicity levels and most were near the drinking water standards (see table 3–D–1). For instance, the maximum cadmium concentration of < 0.01 ppm was observed in 1975; in other years the concentrations were as low as < 0.0001 ppm. The peak concentration of lead, 0.019 ppm, was seen in 1985; the lowest, < 0.005 ppm, occurred in 1982.

The second site, a codisposal facility, contained mostly combined ash (about 76 percent), but also contained household wastes, wastes from a recycling station, and waste–water sludges. This facility accepted waste from 1979 to 1983; the monitoring period covered 1979 to 1985. The DWQI analyzed leachate from this facility for three of the eight EPA–regulated metals. The maximum levels recorded for cadmium and chromium (0.008 ppm and 0.094 ppm, respectively) were close to the highest values at the monofill and were well below the EP toxicity levels. The maximum value for lead, 0.49 ppm, was about 2.5 times higher than the maximum level for the monofill, though still an order of magnitude below the EP toxicity standard.

Malcolm Pirnie, Inc. (1988)

Malcolm Pirnie, Inc. sampled leachate from the Sprout Brook monofill in Westchester County, New York, which disposes of combined ash from the Westchester County Resource Recovery Facility. They took 29 samples from the leachate collection system during the period from October 1986 to November 1987. As shown in the summary of results in table 3–D–2, the values for all metals were well below the EP toxicity limits, though most were somewhat higher than Hjelmar's (1987) results for the monofill.

Table 3–D–1. Concentrations of Metals in Field Leachate from Denmark, 1973–1986

	Monofill	Codisposal	EP Toxicity Limits
Arsenic			
maximum	0.025	—	5.0
minimum	0.005	—	
Cadmium			
maximum	<0.01	0.008	1.0
minimum	<0.0001	0.0006	
Chromium			
maximum	0.08	0.094	5.0
minimum	0.001	<0.003	
Mercury			
maximum	<0.001	—	0.02
minimum	<0.00005	—	
Lead			
maximum	0.019	0.49	5.0
minimum	<0.005	<0.001	
pH			
maximum	10.5	8.7	—
minimum	8.7	6.6	

Note: Values (except pH) given in ppm (mg/l).

Source: Hjelmar, Ole. 1987. "Leachate from Incinerator Ash Disposal Sites." Presented at International Workshop on Municipal Waste Incineration (Montreal, Canada, October 1–2).

Table 3–D–2. Concentrations of Metals in Field Leachate From Westchester County, New York, 1986–1987

	Maximum	Minimum	Average	EP Toxicity Limits
Arsenic	<0.005	<0.005	<0.005	5.0
Barium	0.30	<0.01	0.06	100.0
Cadmium	0.13	<0.001	0.073	1.0
Chromium	0.29	<0.01	0.05	5.0
Lead	0.71	0.005	0.434	5.0
Mercury	<0.0005	<0.0005	<0.0005	0.2
Selenium	0.008	<0.005	0.005	1.0
Silver	0.08	<0.01	0.05	5.0
pH	9.97	7.13	8.41	—

Note: Values (except pH) given in ppm (mg/l).

Source: Malcolm Pirnie, Inc. 1988. Unpublished Data from Westchester County Sprout Brook Ash Residue Monofill, Leachate Characterization Study.

U.S. EPA (1987b)

EPA obtained field samples in late 1986 from three monofills that disposed of combined ash (see table 3–D–3). Samples were taken either from natural seeps or from surface water runoff streams along the bottom of the landfill. Because the facilities either did not have leachate collection systems or because the systems were not functioning prop-

erly, it is not certain that the samples collected were actual leachate (U.S. EPA 1987b, p. 3–35). Facility B had been in operation for 4 or 5 years; Facility C began operations in 1970; no date was given for Facility D. EPA took three grab samples from at least two different locations at each facility. No sample was obtained from a fourth facility (Facility A) because it did not have a leachate collection system and no seeps for gathering a sample could be found.

Table 3–D–3. Concentrations of Metals in Field Samples from Three Monofills

	B	Facility C	D	Overall Average	EP Toxicity Limits
Arsenic	<0.010	0.017	0.027	0.0362	5.0
	<0.010	**0.218**	0.025		
	<0.010	0.013	0.011		
Cadmium	<0.010	<0.010	0.031	0.0162	1.0
	<0.010	**<0.050**	0.023		
	0.044	<0.010	<0.005		
Chromium	<0.005	0.011	0.13	0.1397	5.0
	<0.005	**0.914**	0.099		
	0.024	0.0053	0.069		
Mercury	<0.0002	0.00021	0.00079	0.0014	0.2
	0.0002	**0.0080**	0.00061		
	0.0023	<0.0002	<0.0002		
Lead	0.050	0.068	1.33	0.638	5.0
	0.050	**2.92**	0.925		
	0.206	0.025	0.214		
Selenium	<0.005	<0.025	<0.025	0.0119	1.0
	<0.025	**0.037**	<0.025		
	<0.005	<0.025	<0.005		

Notes: Values given in ppm (mg/l); **bold** indicates the highest values.

Source: U.S. Environmental Protection Agency 1987b. *Characterization of MWC Ashes and Leachates from MSW Landfills, Monofills, and Co–Disposal Sites*, Vol. V. *Characterization of Municipal Waste Combustor Residues* EPA 530–SW–87–028E (Washington, D.C.: EPA, October).

Concentrations for five of six metals analyzed were well below EP toxicity levels. However, some measurements of lead, as with Hjelmar's (1987) and the Malcolm Pirnie (1988) results, were closer to the regulatory limit than the other metals. The second sample taken at Facility C, an older facility, consistently contained the highest concentrations for all six metals. Although no sample in the EPA study surpassed the regulatory limits, two samples, at 2.92 ppm and 1.33 ppm, from two different facili-

ties approached the 5.0–ppm standard although other samples were well below. EPA also took leachate samples from two codisposal landfills. Both facilities disposed of at least 25 percent MWC ash, though only one of the landfills also accepted municipal solid wastes. One landfill began operating in 1985, the other in late 1981. Three samples were collected at each site. Concentrations of cadmium and lead were near or below drinking water levels (see table 3–D–4).

Table 3–D–4. Concentrations of Cadmium and Lead in Field Samples from Two Codisposal Landfills

	Facility 1			Facility 2		
Cadmium	0.011	0.009	0.006	——	——	——
Lead	0.018	0.010	0.022	0.026	0.027	0.018

Notes: Values given in ppm (mg/l).

Source: EPA 1987c, *Characterization of MWC Ashes and Leachates from MSW Landfills, Monofills, and Co–Disposal Sites*, Vol. VI. *Characterization of Leachates from Municipal Waste Disposal Sites and Codisposal Sites*. EPA 530–SW–87–021F (Washington, D.C.: EPA, October).

Sears (1988)

In early 1988, three field leachate samples were taken from a leachate manhole at the Woodburn ash monofill in Marion County, Oregon (Sears 1988) (see table 3–D–5).

These samples show levels of cadmium and lead that are very low—well below the EP toxicity limits—which would be expected from the near neutral–pH level of the leachates.

Denison (1988); Siegler (1989)

Browning–Ferris Industries (BFI) has analyzed field leachates from their ash landfill in Rockingham, Vermont since 1986 (Denison 1988; Siegler 1989). These leachates are probably best characterized as monofill leachates that have been mixed with an unknown quantity of municipal solid waste leachates. At this facility, ash was placed in a lined cell above several feet of municipal solid waste. It is unlikely that the waste or its leachates mixed directly with the ash, but some waste leachate probably mixed with the ash leachate samples.

BFI has reported results from 25 such samples, which include weekly grab and monthly composite samples. About 80 percent of the samples showed concentrations of lead between 0.5 and 2.0 ppm. The only value for lead above 2.0 ppm was a grab sample at 7.1 ppm. In most samples, the pH ranged from 5.5 to 10.7—within the range at which lead typically does not leach. All measurements of cadmium were below EP toxicity levels. These ranged from < 0.01 to 0.489 ppm. The concentration of selenium was measured in about half the samples, at levels ranging from 0.087 to 0.777 ppm. All were below the EP toxicity standard, but three values were above 0.7 ppm, which is near the 1.0–ppm limit.

Table 3–D–5. Concentrations of Cadmium and Lead in Field Leachate from Woodburn Ash Monofill, Marion County, Oregon

	Sample			EP Toxicity Limits
	1	2	3	
Cadmium	bdl[a]	0.0017	0.0011	1.0
Lead	0.011	0.024	0.025	5.0
pH	6.7	6.9	7.2	——

Note: Values (except pH) given in ppm (mg/l).

[a] Below detection limit.

Source: Sears, J. 1988. Marion County Department of Solid Waste, Salem, Oregon. Personal communication, data sheet.

REFERENCES

Bleifuss, Rodney L., Harlan B. Niles, and John Engesser. 1988. "Round Robin Analyses (1) EP Toxicity Test (2) ASTM Water Leach Test." Technical Report NRRI/GMIN–TR–88–06, prepared by the Natural Resources Research Institute, University of Minnesota, for Minnesota Waste–to–Energy Association (Duluth, Minn., January).

Cundari, Kenneth L., and Jeffrey M. Lauria. 1986. "The Laboratory Evaluation of Expected Leachate Quality from a Resource Recovery Ashfill." Prepared for Triangle Conference on Environmental Technology (Chapel Hill, N.C.).

Cundari, Kenneth L., and Jeffrey M. Lauria. 1987. "Resource Recovery Ash and Ashfill Leachate: Westchester County, NY Case Study." Prepared for Air Pollution Control Association 80th Annual Meeting (New York, N.Y., June 21–26).

Denison, Richard. 1988. "Leachate Lead and Cadmium from the BFI Landfill in Rockingham, VT," Unpublished Report (Washington, D.C.: Environmental Defense Fund).

DiPietro, John V., M. Robin Collins, Marcel Guay, and T. Taylor Eighmy. 1989. "Evaluation of pH Oxidation—Reduction Potential on Solid Waste Incinerator Residues." *Proceedings of the International Conference on Municipal Waste Combustion*, Hollywood, Florida, April 11–14.

Eighmy, T. Taylor, Nancy E. Kinner, and Thomas P. Ballestero. 1988. *Codisposal of Lamprey Regional Solid Waste Cooperative Incinerator Bottom Ash and Somersworth Wastewater Sludges: Final Report*. Prepared for Lamprey Regional Solid Waste Cooperative (Durham, New Hampshire, January).

Francis, C.W. 1984. *Leaching Characteristics of Resource Recovery Ash in Municipal Waste Landfills*. Oak Ridge National Laboratory Sciences Division Publ. No. 2456. Prepared for California Waste Management Board (Oak Ridge, Tenn.: ORNL, December).

Francis, C. W., and M. P. Maskarinec. 1987. *Leaching of Metals from Alkaline Wastes by Municipal Waste Leachate*, Environmental Sciences Division Publ. No. 2846. Prepared for the U.S. Environmental Protection Agency (Washington, D.C., March).

Francis, C. W. and G. H. White. 1987. "Leaching of Toxic Metals from Incinerator Ashes," *Water Pollut. Control Fed.* vol. 59, no. 11, pp. 979–986.

Hjelmar, Ole. 1987. "Leachate from Incinerator Ash Disposal Sites." Presented at International Workshop on Municipal Waste Incineration (Montreal, Canada, October 1–2).

Jirka, A., M. Shannon, J. Morris, and P. Parikh. 1987. "Factors Affecting EP Toxicity Metals Results," *Proceedings of EPA symposium on Solid Waste Testing and Quality Assurance*, Vol. I (Washington, D.C.).

Karasek, F. W., G. M. Charbonneau, G.J. Reuel, and H.Y. Tong. 1987. "Determination of Organic Compounds Leached from Municipal Incinerator Fly Ash By Water at Different pH levels." *Anal. Chem.* vol. 59, pp. 1027–1031.

Knudson, James C. 1986. *Study of Municipal Incineration Residue and its Designation as a Dangerous Waste*. Washington State, Department of Ecology, Solid Waste Section (Olympia, Wash.: August).

Malcolm Pirnie, Inc. 1988. Unpublished data from Westchester County Sprout Brook Ash Residue Monofill, Leachate Characterization Study.

Mika, S., and W. A. Feder. 1985. *RESCO Incinerator Residue Research Program: Results, Evaluations, and Recommendations: Final Report*. Prepared for Refuse–Energy Systems Company (Waltham, Mass.).

New York State Department of Environmental Conservation. 1987. *Ash Residue Characterization Project: Summary Report* (Albany, N.Y.: Division of Solid and Hazardous Waste, July).

Ogden Projects, Inc. 1988. *Environmental Test Report No. 179*, regarding Bristol Resource Recovery Facility Residue. Prepared for Ogden Martin Systems of Bristol, Inc. (Fairfield, N.J.).

RAI (Resource Analysts, Inc.). 1987. Unpublished Data from Claremont, N.H., SES Resource Recovery Facility.

Sears, J. 1988. Marion County Department of Solid Waste, Salem, Oregon. Personal communication, data sheet.

Shinn, Claude E. 1987a. *Extraction Procedure Toxicity (EP-tox) Characterization of Municipal Incinerator Ash from Ogden Martin, Brooks* (Portland, Ore.: Department of Environmental Quality, Laboratory Division, May).

Shinn, Claude E. 1987b. *Toxicity Characteristic Leaching Procedure (TCLP), Extraction Procedure Toxicity (EPtox), and Deionized Water Leaching Characteristics of Lead from Municipal Waste Incinerator Ash*. (Portland, Ore.: Department of Environmental Quality, Laboratories and Applied Research Division, July).

Siegler, Ted. 1989. Letter to Michael Gough and Alyce Ujihara, DMS Environmental Services, Inc., Ascutney, Vermont, March 30.

Surgi, Rene. 1986. *Residues from Resource Recovery Facilities: Current Research*. Prepared for Signal Environmental Systems at Marine Sciences Research Center, State University of New York (Stony Brook, N.Y., May 23).

Thomas, L. M. 1988. Statement before the Subcommittee on Transportation, Tourism, and Hazardous Materials, of the House Committee on Energy and Commerce, April 13.

U.S. EPA (U.S. Environmental Protection Agency). 1982. *Test Methods for Evaluating Solid Waste Physical/Chemical Methods*. SW–846 Second Edition (Washington, D.C.: U.S. EPA, July).

U.S. EPA (U.S. Environmental Protection Agency). 1984. "Solid Waste Leaching Procedure Manual: Technical Resource Document for Public Comment SW–924," Draft, (Washington, D.C.: Office of Solid Waste and Emergency Response, March).

U.S. EPA (U.S. Environmental Protection Agency). 1987a. *Characterization of MWC Ashes and Leachates from MSW Landfills, Monofills, and Co–Disposal Sites, Vol. I Summary.* EPA 530–SW–87–028A (Washington, D.C.: EPA, October).

U.S. EPA (U.S. Environmental Protection Agency). 1987b. *Characterization of MWC Ashes and Leachates from MSW Landfills, Monofills, and Co–Disposal Sites, Vol. V. Characterization of Municipal Waste Combustor Residues.* EPA 530–SW–87–028E (Washington, D.C.: EPA, October).

U.S. EPA (U.S. Environmental Protection Agency). 1987c. *Characterization of MWC Ashes and Leachates from MSW Landfills, Monofills, and Co–Disposal Sites, Vol. VI. Characterization of Leachates from Municipal Waste Disposal Sites and Co–disposal Sites.* EPA 530–SW–87–021F (Washington, D.C.: EPA, October).

U.S. EPA (U.S. Environmental Protection Agency). 1987d. *Municipal Waste Combustion Study: Report to Congress.* EPA/530–SW–87–021A (Washington, D.C.).

U.S. EPA (U.S. Environmental Protection Agency). 1989. *Characterization of Products Containing Lead and Cadmium in Municipal Solid Waste in the United States, 1970 to 2000*, prepared by Franklin Associates, Ltd. (Washington, D.C., January).

Visalli, J. R. 1987. "A Comparison of Dioxin, Furan, and Combustion Gas Data from Test Programs at Three MSW Incinerators." *J. Air Pollut. Contr. Assoc.* vol. 37, pp. 1451–1463.

Wakimoto, T., and R. Tatsukawa. 1985. "Polychlorinated Dibenzo–p–dioxins and Debenzofurans in Fly Ash and Cinders Collected from Several Municipal Incinerators in Japan," *Environ. Health Perspec.* vol. 59, pp. 159–162.

4

Research: Source Reduction, Disposal, Treatment and Reuse

The nation confronts difficult decisions about municipal solid waste management, and it will continue to do so for some time. Decisions would be improved by a systematic comparison of the risks, costs, and benefits of landfilling, incineration, recycling, waste reduction, and other options for the management of municipal solid wastes. Such research would be costly and time consuming, and deciding whether or not to do it would be an important decision for industry or government. Until it is done, however, we will be "flying blind" to some extent.

In the meantime, specific research to improve ash management could reduce risks from ash and enhance discussions about how it should be managed. Because metals are the substances most likely to cause harm from ash, they have been the focus of most research to date. There are four generic ways to reduce these risks. First, the metals can be removed from the waste stream prior to combustion; this is one type of source reduction. Second, ash can be disposed of in properly designed landfills so that the metals do not reach groundwater in unacceptable quantities. Third, ash can be treated to remove metals before disposal or else treated so that the metals are not released when the ash is placed in a landfill. Finally, the ash can be reused so that humans are not exposed to the metals and so that it does not have to be landfilled.

SOURCE REDUCTION

One means of reducing risks from ash is to remove items that are the source of leachable toxic constitutents from the waste stream of incinerators. This step should reduce the toxicity of the ash leachates. Lead and cadmium stand out as concerns in consideration of municipal waste combustor (MWC) ash, and identifying the sources of leachable lead and cadmium in the waste stream is an important step in reducing that concern. According to a study by Franklin Associates for EPA (U.S. EPA 1989a), nearly 98 percent of the lead in municipal solid waste comes from noncombustible sources such as lead–acid batteries (65 percent of the total) and consumer electronic devices (27 percent). Combustible sources of lead include plastics and pigments used in lacquers and paper. The main sources of cadmium are household batteries (52 percent) and plastics (28 percent). Not all of these sources may contribute equally to the leachable metals that are detectable in the EP tox test. Many MWC operators try to screen lead–acid batteries out of the trash before it is collected or put into the incinerator; it is possible, however, that the lead in pigments and plastics is more leachable than the lead in batteries.

Sommer et al. (1988) describe results of a study that demonstrates that sorting of municipal solid waste before incineration may reduce weight and toxicity of ash. A presort method was developed to remove aluminum metals, ferrous metals, batteries, and glass/grit from municipal solid waste beforehand. Aluminum metals, heavy ferrous metals, and automobile batteries are sold for recycling, glass/grit, and tinplate are disposed of in landfills. The presorting step reduced ash weight by about 50 percent. EP tox test results on the bottom ash before and after the presorting step showed some reduction in the concentration of lead, although the level of cadmium rose slightly (see table 4–1). The authors also reported improved combustion performance, sometimes called "enhanced fuel characteristics" with presorted fuel and suggested that the considerable advantages—improved combusion performance, recycling possibilities, reduced ash production, and possible reduced ash toxicity—may provide the "necessary incentive to incorporate materials recovery with mass burn systems" (Sommer et al. 1988, p. 15).

Table 4–1. EP Tox Test Results from Bottom Ash Before and After Presorting the Municipal Solid Waste

| Metals | Facility Location | Leachate Concentrations | |
		Before Presort	After Presort
Lead	Nashville	2.42	0.69
	Salem	15.20	12.30
Cadmium	Nashville	0.10	0.42
	Salem	0.21	0.28

Note: Values given in ppm (mg/l).

Source: Sommer, E. J. Jr., G. R. Kenny, J. A. Kearley, and C. E. Roos. 1988. "Emissions, Heavy Metals, Boiler Efficiency, and Disposal Capacity for Mass Burn Incineration with a Presorted Municipal Solid Waste Fuel," Paper presented at Air Pollution Control Association Meeting, June 20–24, 1988, Dallas, Texas.

Further studies are needed to decide if removal of other items from municipal solid waste can reduce the toxicity of ash, and other research and development will be necessary to devise methods to remove those items and to decide upon how to recycle or dispose of them. It is likely that such methods will involve costs and perhaps risks from recycling and disposal, and those factors will also have to be considered.

DISPOSAL

In addition to reducing toxicity, ash management, as with the management of all wastes, can be improved by the development of better disposal methods. Projects that lead to improvements in siting and design of landfills, to methods that reduce the entry of water into landfills, and to techniques for field sampling and measurement of leachates will all be applicable to ash monofills as well as to other landfills.

Guidelines for Ash Disposal

In 1988, the U.S. EPA (1988) completed draft guidelines for the disposal of MWC ash; the guidelines were never officially released but were widely distributed. These guidelines recommended different facility designs for different types of ash. For bottom or combined ash monofills, they recommended a composite liner consisting of a layer of packed, low–permeability clay overlaid with a flexible synthetic liner, and a leachate collection system located above the synthetic liner. A more stringent containment system, recommended for codisposal sites, consists of a double–liner system, including a second flexible synthetic liner and a second leachate collection system above the composite liner. EPA recommended the same double–liner system for disposal of fly ash. Recommendations from Waste Management, Inc. (undated)—a large waste collection and disposal company—generally parallel those of EPA for ash monofills and ash codisposal sites.

Both EPA and Waste Management, Inc., recommend that ash disposal landfills be constructed to minimize the amount of rainwater (1) entering the facility where it can cause leaching and (2) running off the facility where it can contaminate water supplies. In addition, Waste Management, Inc., recommends a final cover of hard–packed earth or a combination of hard–packed earth overlaid with a synthetic sheet. Topsoil is to be placed on top of the hard–packed earth or synthetic material, and plants are to be grown in the topsoil. The purpose of the top cover is to reduce the amount of water that enters the landfill.

Legislation introduced into the 100th and the 101st Congress specified landfill designs that are similar to those laid out by EPA and Waste Management, Inc. (discussed further in Chapter 5). Some observers, in particular the Environmental Defense Fund (EDF), argue that all ash disposal sites, including monofills, should have double composite liners. Double composite liner systems consist of a layer of hard–packed clay, a synthetic liner, a leachate collection system, then another layer of clay, synthetic liner, and leachate collection system.

Changes in Ash pH, Chemistry, and Permeability

The exact fate of materials that are placed in landfills is not known. Efforts at containment and at treatment of leachate can be made without that knowledge, but knowing what conditions favor or inhibit the liberation of metals from ash could focus research on the most appropriate containment and treatment methods. For instance, the pH of ash may change as rain or acid rain washes through the ash. If the pH decreases into the acid range, the solubility of metals may increase. However, that effect might be offset because the sulfates present in acid rain can form relatively insoluble sulfur compounds with heavy metals, which may not be leached at the pHs associated with acid rain.

The importance of changes over time is illustrated by the results of Mika and Feder (1985). New ash was added to an uncovered ashpile for 7 years. When a boring was made into the 45–foot–high pile, ash from most levels except the top 2 feet consistently passed the EP tox test. Apparently, changes over time reduced leachability of metals. Some of the metals had leached from the ash into the underlying soil, but measured concentrations of metals in water under or near the ashpile never exceeded the limits established for drinking water.

In addition, ash that contains lime can "set up" into a concretelike matrix, which probably affects permeability and leaching. Furthermore, Pherson (1988) cites studies done by Barnes et al. (1985) which demonstrate that addition of a mixture of lime and fly ash to a lead solution reduced the solubility of the lead. This finding is consistent with the idea that the addition of lime may immobilize lead through chemical reactions despite the high pH of the lime. The consequences of the setting up of lime–containing ash are being investigated, and the results of those studies may have important consequences for ash landfill design.

Unfortunately, there may also be a disadvantage to adding lime to ash, because the high pH may adversely affect clay liners (U.S. EPA 1988). To investigate this problem, EPA (Wiles 1989) plans to review the existing literature about the properties of clay liners to determine if additional information is necessary to understand clay–alkaline interactions. If more tests are desirable, EPA would then institute the appropriate research program.

Additional measurements of pH, chemistry, and permeability under different environmental conditions would be invaluable for assessing the accuracy of the models that have been developed to predict the behavior of metals in ash monofills. Equally important, those measurements would provide real–world data about the liberation of metals. State regulatory agencies as well as EPA will be among the ultimate users of such data. For that reason, collection and analysis of data about pH and other chemical characteristics will be more useful and acceptable if sampling and measuring techniques are jointly agreed upon by MWC owners/operators and regulators.

An example of a joint venture in data collection concerning ash is already under way. The EPA and the Coalition on Resource Recovery and the Environment (CORRE), a consortium of municipal governments, incinerator vendors, and other manufacturers, are collaborating on a jointly funded project to gather information "about municipal incinerator ash, ash handling and management procedures, ash leach test behavior, and leachate properties at actual ash disposal sites" (Shaub 1988). The study will involve use of only EPA–approved sampling and analytical methods as well as quality control measures. When complete, it will be peer reviewed before public release. Should this effort be successful in producing data that are informative and useful, it could provide a model for other collaborative efforts between the private and public sectors.

Alternatives to Landfills for Disposal

The Swedish National Environmental Protection Board has decided against liners and leachate collection systems for ash disposal. Instead, the board recommends locating unlined ash monofills near oceans or other large bodies of water. The ash monofills are covered with soil and vegetation to reduce the amount of water that enters them, but whatever water from rain and snow moves through the ash is allowed to move out of the unlined ash monofills and into the ocean or other large body of water where metals and salt will be diluted to acceptable concentrations (Modig 1988). In the states of Bavaria and Hesse in West Germany, fly ash is disposed of in salt mines where the salt content of ash will certainly have no effect and where the possibility of water infiltration is remote (Linnerooth 1988).

Such disposal solutions illustrate the possibility of exploiting local conditions in the development of disposal methods. Legislation so far considered by the U.S. Congress would have allowed either the EPA or state agencies to consider deviations from the requirements for liners and leachate collection if an equal degree of environmental protection could be achieved through alternative means. Different pieces of legislation varied, however, in how much consideration could be given to local conditions. According to some experts, waste management companies generally favor consideration and exploitation of local characteristics as alternatives to prescribed disposal methods, whereas environmental groups do not. The latter fear that local decison-

makers may overuse consideration of local conditions to the point that health and environmental protection could be compromised.

TREATMENT OF ASH

"Treatment" is used here to refer to methods to remove toxic substances from ash and to methods to stabilize or solidify ash so that toxic substances in ash become nonreactive and immobile to one degree or another.

Removal of Toxic Substances from Ash

One method that might be used to treat ash is to leach metals from it by passing ash through acidified water prior to landfilling. The metals could then be extracted from the water and recycled or disposed of, and the water reused for extraction. This method has the advantage of not adding to the volume of ash to be landfilled, and it removes the hazardous substances rather than confining them in the ash. Some research is being undertaken to examine the possibilities of this method (Hasselriis 1988a). Among the problems to be resolved are optimal ratios of water, acidity, and ash, and the extent to which leachable metals can be removed by treatment.

Solidification and Stabilization Methods to Inhibit or Prevent Leaching

Pherson (1988) reviewed eight stabilization and solidification systems that have been used for one waste or another, and discussed four that have been used or might be used for MWC ash. Three methods involve mixing ash with other substances. Two have been used already: mixing the ash with either (1) Portland cement or (2) lime and pozzolan (materials that are added to cements). The third possibility is to mix the ash with hot asphalt or bitumen.

Pherson (1988) examined two proprietary lime/pozzolan products as well as Portland cement for their capacity to reduce leaching of lead and cadmium from combined ash that contained lime scrubber residue. The ash, as received from a

MWC, had solidified and had to be broken up with a hammer and chisel before it could be crushed for the experiments. Then weighed amounts of ash were mixed with water and measured amounts of one of the solidification materials and allowed to cure for 3 days. At that time the solids were crushed, ground up, and tested with the EP tox test. Ash treated with either 1 part lime/pozzolan product to 10 parts ash or 1 part Portland cement to 10 parts ash passed the EP tox test. As will be seen below, cement and lime products are both offered in ash treatment systems.

Combined ash that contains fly ash with high levels of lime from scrubbers can have a pH of 12 or more. When exposed to water, it undergoes a chemical reaction and sets up as concrete (Goodwin 1988). Calibrated addition of water and lime and application of pressure to the ash can produce a stronger concrete. A test is being carried out at Milbury, Massachusetts, in which ash is being treated with additional lime and water and rolled with a highway construction roller. The resulting concrete area, about the size of a football field, will be examined as it undergoes the usual freeze–thaw and weathering cycles to see how its properties change over time. According to Goodwin (1988), the concrete made in this way costs $4.00 per ton for chemicals and has a water permeability of less than 10^{-7} centimeters per second, which is equivalent to that required for clay bottom liners in hazardous waste landfills. Furthermore, the rolling of the ash reduces its volume by about 50 percent, meaning that landfill space fills more slowly. Although further tests over longer time periods are necessary, available results indicate that metals are bound up in the concrete and are not released at levels exceeding EP toxicity limits. In fact, the EP tox test, when applied to crushed concrete made as part of the disposal project, revealed that leached concentrations were 100 times lower than from the untreated ash (Goodwin 1988).

Another possible treatment involves heating municipal solid waste ash in an electric arc furnace at temperatures sufficient to vitrify the ash (turn it into glass). This process produces two products:

one is a glasslike frit, which resembles a dark or black volcanic stone called obsidian, that is about 3 times the density of ash. It is completely impermeable to water, and materials within it are not leachable. The other product is a metal ingot that is 10 times more dense than ash. The increased densities mean, of course, corresponding decreases in volume and reduced demand for landfill space. There is no shortage of power for the electric arc furnace at a waste–to–energy incinerator, but the large amount of energy required is estimated to cost from $77 to $150 per ton. Japan is now using this process on one incinerator (Hasselriis 1988b), and New York State is attempting to organize financial supporters for a test of the process (Fiesinger 1987).

Several member firms of the Hazardous Waste Treatment Council offer fixation—chemical treatment of wastes to immobilize toxic substances—to their customers. For instance, one firm markets a combination of chemicals that can be mixed with ash in equipment as simple as a cement mixer. The addition of chemicals and the mixing can be performed by incinerator staff without expensive training or equipment at an estimated cost of $20 to $35 per ton (or less), and according to a company spokesman (Belisle 1988), every sample of ash that the company has treated has passed the EP tox test. The product of the treatment is a pea–size gravel which can be disposed of in any nonhazardous waste site because it possesses no hazardous characteristics. Another alternative, as described in the next section, is to use it as roadbed material and for other applications that call for gravel.

Another company has prepared a mixture of chemicals that are now used at one MWC and will be used at another in the near future. The cost of treatment is less than $5 per ton and the treated ash can be disposed of as a nonhazardous waste. The company has not decided whether to pursue this market.

Davidovits et al. (1988) have examined ancient concretes made by the Egyptians and Greeks, deduced how they were made, and applied those techniques to treating hazardous wastes. The process essentially turns waste into "rock" and greatly reduces the amount of metals that can be leached from it. It has been applied to mining and other wastes but not yet to MWC ash (Comrie 1988a; Davidovits et al. 1988).

EPA has announced a program that will attempt to provide "credible data [on the effectiveness of solidification/stabilization and other proposed MWC ash treatment techniques] that can be used by municipalities to plan and implement ash management strategies" (U.S. EPA 1989b). Vendors of those techniques have been invited to treat ash samples at "an EPA supported facility," or, if that is impossible, "at the vendor's facility, [where] EPA technical staff or representatives will observe the process demonstration." Either bottom ash, fly ash, fly ash with scrubber residue (scrubbers are lime-containing air pollution control devices), or combined ash with scrubber residue can be treated, and the EPA will pay for the completion of several tests that will be applied to the untreated and treated ash samples to determine the success of the various techniques. EPA estimates that 12 to 18 months will be necessary to complete the tests and evaluations once the process demonstrations begin. Therefore, if the demonstrations begin at the end of 1989, results from this joint EPA–private sector project will likely be available in 1991.

Incentives/Disincentives for Treatment

So long as there is any possibility that ash might someday be regulated as a hazardous waste, treatment methods that render the ash nonhazardous and that are approved by EPA might find ready markets. Even in the absence of such approval, methods that are relatively inexpensive and that ensure that treated ash would pass the EP tox test (or its equivalent or replacement) might be purchased by owners/operators. Such treatments might be regarded as affordable insurance against the possibility that untreated ash disposed of as a nonhazardous waste might be reclassified.

Most MWC vendors are believed to be performing research on methods to treat ash to reduce leaching of metals. However, vendors are reluctant to discuss that research because of the intense competition for contracts for new MWCs. Treatment will probably increase the cost of ash disposal directly and indirectly. The direct increases come from the cost of chemical additives and increased handling, and the indirect increases result from the increased volume of ash, which requires additional landfill space. No vendor wants to add an item, such as treatment, that bears additional costs to the package of hardware and services offered to prospective customers. Furthermore, disposal technologies that rely on conventional technologies involving liners, leachate collection, and monitoring wells will probably be in place before a regulatory system is in place to validate treatment methods. Thus, it is probably more straightforward to prepare for conventional containment than to develop and then seek approval for a treatment technique.

Overall, there appears to be little incentive for owners/operators to purchase treatment methods. Both the lack of an approved process as well as the uncertain nature of ash classification probably contribute to the limited use of treatment methods. If Congress does enact legislation that excludes ash from classification as a hazardous waste, the costs of treatment could be seen as unnecessary expenditures. Furthermore, to treat ash, if it can be disposed of as a nonhazardous waste, could appear to be a tacit admission that untreated ash is hazardous or, at least, more toxic than other nonhazardous wastes. On the other hand, legislation considered by Congress has contained language that would allow disposal of treated ash in Subtitle D landfills. If such a provision becomes law, it could provide an important incentive for treatment. As with so many aspects of the ash issue, interactions between research and policy decisions will decide whether or not treatment of ash will become standard practice.

REUSE OF ASH

Ash contains metals and grit and glass. Fiesinger and Zoto (1987) estimated that ash contains 40 to 50 percent glass; 10 to 30 percent ferrous metals; 1 to 5 percent nonferrous metals; 1 to 2 percent ceramics, stones, and bricks; 0.5 to 12 percent partly burned organics; and 15 to 30 percent pure ash. The metals in ash clearly are valuable, and the bulk of the heterogeneous mixture of glass and nonburnable minerals in ash also may have value as a building material. To the extent that ash can be reused, landfill demand will be reduced, and to the extent that a market can be found for it, any profit from selling ash might go to offset tipping fees for municipal solid waste.[1]

Reuse of Specific Materials

Somewhat more attention is being given to recovery of recyclable articles before, rather than after, the incineration step. But recovery is also possible after incineration. The incineration process reduces the volume of material by as much as 90 percent, and most organic materials have been burned away. The removal of organic materials is an advantage because plastics, paper, and other burnable items do not contaminate ferrous metals that are recoverable after incineration. For instance, ferrous metals are routinely separated from ash before the ash is transported to a landfill from the MWC in Westchester County, New York (Cundari and Lauria 1986).

Mahoney (1986) describes three incinerator facilities at which both ferrous and nonferrous metals are recovered after incineration. In addition to generating some revenue and making small reductions in the ash volume, the removal of aluminum may have an additional benefit for ash reuse. The material that remains after metal removal can be used as aggregate (like sand or gravel) in various construction materials, but the presence of aluminum in ash reduces the strength of concrete made from it.

Therefore removal of that metal may increase the use of ash as aggregate. Ferrous metals and aluminum are recovered at, respectively, 98 and 97 percent purity in the Mitsubishi Ash Sintering process (Matsuda 1980).

Reuse in Building Materials

Goodwin and Schuetzenduebel (1987) discuss the chemical composition of MWC ash and point out that it, like ash from coal–fired power plants, resembles Portland cement. A number of investigators have examined the use of MWC ash in fabrication of cement–based products, but consideration of such uses must include recognition of the toxic substances that are present in MWC ash and absent from Portland cement.

The SEMASS incinerator at Rochester, Massachusetts, uses methods similar to those described by Mahoney (1986) to separate ferrous and nonferrous metals from ash (Mahoney and Mullen 1988). That facility also keeps bottom ash separate from fly ash and, prepares a material from bottom ash, for use in concrete manufacture. Combustion of 570 tons of refuse–derived fuel at SEMASS each day produces 120 tons of bottom ash. One ton of that ash is separated out as nonferrous metal, 7 tons as ferrous metal, and 1 ton as oversized material (greater than 4–inch size), leaving 112 tons of material that is "similar, in color and texture, to Upstate New York gravel" (Mahoney 1986), and which can be used in fabrication of concrete. Concrete blocks made with this material were initially 60 percent stronger in compression tests than those made with conventional aggregates, and compression strength increased over time (Mahoney and Mullen 1988). Some of the bottom ash, when tested with the EP tox test, exceeded the critical level for lead, but none of the concrete blocks fabricated from ash exceeded the EP toxicity limit. Further, when the blocks were ground to pieces smaller than 9.5 millimeters, the pieces also passed the EP tox test. Mahoney and Mullen (1988) suggest that the metals from the ash are chemically immobilized in the concrete as well as being physically encapsulated.

---

[1] Tipping fees are the prices charged by a landfill or incinerator owner/operator for its services.

More innovative uses of the boiler aggregate concrete blocks involved grinding and surfacing faces of the blocks to accentuate the colored glass particles that they contain (Mahoney 1986). Development of a commercial market for such materials would, of course, reduce the amount of bottom ash that would require landfilling.

Denmark (Danish Ministry of the Environment 1983) permits the use of ash that contains less than 25 percent fly ash as fill or base for walkways, roads, and yards that are to be covered with asphalt or concrete and for laying pipe. Ash must not be used within 20 meters of "waterworks" (wells) that supply drinking water, laid below the groundwater table, nor used in thicknesses that, on average, exceed 1 meter. Rigo (1982) reported on the results of testing storm–water runoff from the surface of a Danish parking lot as well as the water that he reported to have run through the parking lot's subbase, which was made of MWC ash. The water from the surface was more contaminated with hazardous materials, which had been deposited on the parking lot from auto exhausts and drips.

In addition to being used as a base for roads, ash has also been used as a component in asphalt paving. Such asphalt passes all tests for durability, compression strength, and other properties required of asphalt, and it has been used on a few roads in this country. Despite these results, additional tests about the leachability of toxic materials from ash used in road building as well as more tests of the performance of products that incorporate ash will be required before ash finds a large market.

According to Fiesinger and Zoto (1987), bottom ash is widely used as fill and in road building in Europe and Japan, and Hasselriis and Aleshin (1986) report that the U.S. Department of Transportation has concluded that ash is suitable for use as structural fill and as aggregate for asphalt and concrete. The State of New York has also approved aggregate, such as that produced by the SEMASS incinerator, for use as a landfill cover (Mahoney 1986).

Processes have been developed that make use of ash from incineration processes in the manufacture of construction materials. Although no materials made from MWC ash have yet been marketed, two materials made from steel plant ash have been. According to a company spokesman, the process effectively ties up heavy metals within a polymerized matrix (Comrie 1988b).

Researchers at the State University of New York at Stony Brook are investigating use of ash in making concrete to be used in construction of artificial reefs as well as structures on land. In the first step of the process, the ash is mixed with either hydrated lime or Portland cement and fabricated into hollow bricks of the type familiar to everyone who has seen a "bricks and boards" bookshelf.

Bricks made with Portland cement and MWC ash are stronger than those made with hydrated lime and ash. They have about 80 percent of the compressive strength of bricks made with Portland cement and natural aggregate, making them sufficiently strong for ocean disposal. Comparison of the compressive strength of blocks made with conventional aggregate and MWC ash were made after 380 days of immersion in saltwater. The strength of the conventional blocks decreased, but that of the MWC ash blocks increased. All metals, including lead and cadmium were effectively retained within the blocks after 380 days of immersion, the longest time yet reported. Furthermore, examination of sessile animals that attached themselves to the blocks showed that there were no differences in metals concentrations between those that grew on conventional blocks and those that grew on MWC ash blocks (Roethel and Breslin 1988).

All of these efforts at reuse of ash are being developed by the private sector or by university–based researchers. To the extent that these products find markets, they can reduce the disposal problem. However, several hurdles must be overcome to find markets. Ash for construction use must compete with already cheap materials such as sand and gravel, and only if ash is sufficiently less expensive than those materials will it be chosen over them. On the other hand, MWC operators may be willing to pay materials manufacturers to take the ash away, which would provide a financial incentive for its use. Furthermore, the proximity of some MWCs to industrial sites might result in reduced transportation costs as compared to the cost of transporting sand and gravel for construction materials manufacturing (Penner 1988) and perhaps other uses.

Senator Quentin N. Burdick (D–North Dakota) introduced a bill (S. 196) early in the 101st Congress that states that 3 years after enactment, unless the EPA Administrator imposes other regulations, no one will be allowed to reuse ash unless the ash is treated and leachates from an extraction–type test such as the EP tox test do not surpass drinking water standards. In addition, an entire section of the bill discusses reuse which, if the bill is enacted, would demonstrate congressional interest in reuse.

FUTURE RESEARCH

A representative of one company that has developed a method to treat ash stated that there is no incentive for treating ash as long as it appears likely that ash will be disposable as a nonhazardous waste. At the same time, the uncertain regulatory fate of ash is probably hindering its reuse. If there is any chance that ash might be declared a hazardous waste, its reuse might be severely restricted even if products fabricated from ash passed all toxicity tests. These comments illuminate the intertwined nature of policy and technology in the MWC issue. Research and development, and, to a greater extent, application of new methods await decisions about how ash will be regulated.

Although legislation considered so far would classify ash as a nonhazardous waste, most has included provisions that would encourage treatment. For instance, legislation would allow ash to be co-disposed of in regular Subtitle D landfills if certain criteria based on tests are met, and most likely the criteria could be met through some form of treatment. Furthermore, treatment may be necessary for ash that is to be reused.

The research itself can probably be accomplished within a relatively short time once it is de-cided which projects to pursue. For example, a committee of experts assembled at the National Science Foundation concluded:

> Because of the practical urgency of the issue of ash stabilization and disposal, current industry efforts may be expected to yield acceptable solutions within a few years (Penner 1988, p. 11).

Nevertheless, until regulatory uncertainties are resolved and policy decisions are made about what research is most important, research will probably move only slowly.

SUMMARY

Research has shown much promise for methods to treat ash to reduce the leachability of toxic substances and to reuse it in road fill and concrete. More general research directed at targeting toxic substances for removal from the waste stream and improving landfills will also benefit ash management.

Like all aspects of the management of MWC ash, research and development efforts for ash treatment and reuse will be guided by decisions made by Congress and EPA. The current regulatory uncertainty that surrounds ash probably hinders research. But legislation that has been considered by Congress could clarify the situation. Some bills contain provisions that would allow disposal of MWC ash in Subtitle D facilities if the ash passes certain tests, and treated ash might pass such tests more often.

REFERENCES

Barnes, D., P. A. FitzGerald, and R. McFarland. 1985. "Chemical Fixation for the Management of Heavy Metal Wastes," *Chem. Eng. Australia* vol. 10, pp. 19–32.

Belisle, H. 1988. Lopat Enterprises, Inc. Personal communication (telephone conversation), October 21.

Comrie, D. C. 1988a. "New Hope for Toxic Waste." *The World and I.* (August 1988), pp. 171–177.

Comrie, D.C. 1988b. Comrie Waste Management Consultants. Personal communication (telephone conversation), September 30.

Cundari, K. L. and J. M. Lauria. 1986. "Ash Monofills and Leachate," *Waste Age* vol. 17, no. 11 (November) pp. 82–88.

Danish Ministry of the Environment. 1983. Order Governing Utilization of Slag and Flyash. Order No. 568 of December 6, 1983. [Translated by Helen Hasselriis].

Davidovits, J., D. C. Comrie, J. H. Paterson, and D. J. Ritcey. 1988. "Applications of Geopolymer Technology in the Prevention of Groundwater Contamination." Paper presented at Haztech Canada 2nd Annual Hazardous Materials Management Conference, Mississauga, Ontario, May 10–12.

Fiesinger, T. 1987. Memorandum: Ash Management Research. Albany, N.Y.: New York Energy Research and Development Authority.

Fiesinger, T. and G. Zoto. 1987. "Incinerator Ash Management Project." Final draft. Albany, N.Y.: New York Energy Research and Development Authority.

Goodwin, R. W. 1988. Consulting engineer. Personal communication (telephone conversation) September 29.

Goodwin, R. W. and W. G. Schuetzenduebel. 1987. "Residues from Mass Burn Systems: Testing, Disposal, and Utilization Issues. Paper presented at New York State Legislative Commission of Solid Waste Management and Materials Policy Conference, New York City, February 11–14.

Hasselriis, F. 1988a. Gershman, Brickner & Bratton, Inc. Oral presentation at meeting sponsored by CORRE (Washington, D.C. November 2, 1988).

Hasselriis, F. 1988b. Gershman, Brickner & Bratton, Inc. Personal communication (telephone conversation), September 30, 1988.

Hasselriis, F. and E. Aleshin. 1986. "How Residues from Waste to Energy Plants Can be Used Safely." Paper presented at Association of State and Territorial Solid Waste Management Officials Conference, Los Angeles, Calif., September 17–19.

Linnerooth, J. 1988. Institute for Applied Systems Analysis, Laxenburg, Austria. Letter to Allen Kneese, Resources for the Future. October 6.

Mahoney, P. F. 1986. "There's Gold in That There Ash!" Waste Age (April) pp. 265–270.

Mahoney, P. F. and J. F. Mullen. 1988. "Utilization of Ash Products from Combustion of Shredded Solid Waste." Paper presented at Air Pollution Control Association Annual Meeting, Dallas, Tex., June 20–24.

Matsuda, T. 1980. "New Ash Sintering Process for Municipal Refuse Incinerators." Paper presented at Third Recycling World Conference, Basel, Switzerland, September 29–October 1.

Mika, J. S. and W. A. Feder. 1985. RESCO Incinerator Residue Research Program: Results, Evaluations and Recommendations: Final Report. Prepared for Refuse–Energy Systems Company, Waltham, Mass.

Modig, S. 1988. "Swedish View of the Ash Issue." Paper presented at a seminar on An Environmentally Acceptable Solution to the Solid Waste Problem (Washington, D.C., March 23).

Penner, S. S. 1988. Research Recommendations on Incineration of Municipal–, Hazardous–, and Biomedical Wastes, National Science Foundation workshop (April 18–19) (La Jolla, Calif.: Center of Energy and Combustion Research).

Pherson, P. 1988. "Environmental Engineering Fixation of Lead in Resource Recovery Ash by Solidification." PhD thesis, Purdue University, Lafayette, Inc.

Rigo, H. G. 1982. State of the Knowledge Report on the Disposal of Incinerator Ash (Rigo and Rigo Associates: Duxbury, Mass.).

Roethel, F. J. and V. T. Breslin. 1988. "Interactions of Stabilized Incineration Residue with the Marine Environment." Solid Waste and Power vol. 2, no. 5 (October) pp. 42–48

Shaub, Walter. 1988. "Joint CORRE–US EPA Ash Research Program Initiated," CORRE Newsletter vol. 2, no. 7, p. 5.

Sommer, E. J. Jr., G. R. Kenny, J. A. Kearley, and C. E. Roos. 1988. "Emissions, Heavy Metals, Boiler Efficiency, and Disposal Capacity for Mass Burn Incineration with a Presorted Municipal Solid Waste Fuel." Paper presented at Air Pollution Control Association Meeting, Dallas, Tex., June 20–24.

U.S. EPA (U.S. Environmental Protection Agency). 1988. "Draft Guidance: Municipal Waste Combustion Ash." EPA/530–SW–88–006, OSWER Policy Directive No. 9573.00.1 (Washington, D.C., March).

U.S. EPA (U.S. Environmental Protection Agency). 1989a. Characterization of Products Containing Lead and Cadmium in Municipal Solid Waste in the United States, 1970 to 2000 (Washington, D.C.; EPA, January 1989).

U.S. EPA (U.S. Environmental Protection Agency). 1989b. U.S. EPA Program for Evaluating Solidification/Statilization and Other Technologies for Managing Municipal Waste Combustion Ashes. (A circular distributed to vendors of ash treatment technologies) (Cincinnati, Ohio: EPA).

Waste Management, Inc. Undated. Recommended ash disposal practices: Ash monofills and co–disposal facilities minimum requirements. Photocopied typescript, 3 pages.

Wiles, C. 1989. Clay Liner Issue (description of planned U.S. EPA research, dated Nov. 20, 1987), 11 pp.

5

Solutions to the Dilemmas Presented by MWC Ash

Views on the acceptability of incineration and views on acceptable solutions to the ash problem are closely linked. Those who are opposed to incinerators view preventing construction of new MWCs and the closure of existing ones as a solution to the problems posed by ash. Those who are convinced of the usefulness of incinerators tend to think that managing ash as a hazardous waste is unnecessary and needlessly expensive. And those who see incineration as a potentially important technique for waste management—perhaps the largest group—tend to think that effective solutions can be found for the ash problem but that these solutions remain elusive.

Probably everyone agrees that ash presents some risks, and any solution that retains MWCs as an option in solid waste management will have to acknowledge and provide for the management of those risks. The existing federal system so far has failed to do that. Technical options are available to address the ash problem, some of which may offer considerable potential to reduce the risk posed by ash. Yet, in order for any of these options to be employed, action is required on at least one of three fronts: judicial, legislative, or regulatory.

LEGAL SOLUTIONS

In recent years, public interest groups have used the courts to influence federal policy. In the area of MWC ash, the Environmental Defense Fund (EDF 1988) has brought suits in federal court against the owners and operators of two incinerators. These cases are expected to go to trial in 1990, and a ruling in either one could set a precedent for at least a partial legal "solution" to the ash problem. The basis for the suits is data that show that ash produced from these facilities, when tested using the EP tox test, exceeded the regulatory levels for lead and/or cadmium. EDF argues that the ash should be classified as a hazardous waste and thus the facilities should be regulated under Subtitle C. The lawsuits charge the facility owners and operators with violation of several Subtitle C provisions.

EDF's lawsuit challenges an ambiguous part of the law: the applicability of the household waste exclusion to MWCs. Although this exclusion addresses municipal solid wastes, it is silent on whether the ash residues resulting from these wastes (which often exhibit hazardous characteristics) should dictate whether or not a facility is subject to hazardous waste requirements or exempt from Subtitle C. EDF's position, as quoted from their April 13, 1988 testimony before the House Subcommittee on Transportation, Tourism, and Hazardous Materials, is that the exclusion "does not extend to *ash* generated by the facilities to which it applies". (See Chapter 2 for a more detailed discussion on the household waste exclusion.)

Though a ruling for or against EDF would provide some clarification to the regulatory status of ash, it would not resolve the ash management problem entirely. Unanswered questions would remain. For instance, if EDF prevails and the court finds that the facilities in question do not fall under the household waste exclusion, some incinerators and incinerator ash would clearly be classified as hazardous and become subject to Subtitle C regulations. EPA would be forced to require ash testing and the EP tox test would become the arbiter of a facility's status (hazardous or nonhazardous) for the disposal of its ash.

While application of the strict Subtitle C regulations to ash should allay some fears about risks from ash, using the EP tox test as the basis for an ash management program would be far from optimal. Because of the many problems that have plagued use of the EP tox test as a tool to characterize ash, there is little reason to expect that it can reliably indicate which ash should be classified as hazardous. Furthermore, in our dichotomous system for managing wastes, ash that did not surpass EP tox test levels would fall under Subtitle D, where regulations are currently very weak.

The elimination of the household waste exclusion and its impact on incineration would depend on the quantities of ash that are eventually declared hazardous. Whether there is adequate capacity to dispose of ash as a hazardous waste and how much new capacity will be needed is open to debate. Certainly the higher cost of Subtitle C disposal (see Appendix 5–A) would restrict the use of MWCs as a solid waste management option, in part due to the effects of increased costs, but more because of the stigma that would be associated with a MWC producing a waste that has been declared hazardous. On the other hand, this situation could prompt Congress, if it decides that ash should be managed as a nonhazardous waste, to act more quickly.

If EDF "loses" the suits and the court finds that the household waste exclusion does apply (meaning that ash from MWCs with energy recovery are completely exempt from Subtitle C), ash will be subject to Subtitle D regulations until Congress or EPA takes further action. Of course, EDF's lawsuits have no affect on facilities that do *not* recover energy. Such facilities, which presently handled about

one–third of incinerated municipal solid waste in 1987, were clearly excluded from the household waste exclusion, and they must meet Subtitle C requirements under existing law if leachates from their ash exceeds the EP tox limits.

LEGISLATIVE SOLUTIONS

Congressional Initiatives

During 1988, Congress considered several pieces of legislation that could have resolved the ash controversy. Through November 1989, the 101st Congress had two ash bills under consideration: S. 196 by Senator Burdick and H.R. 2162 by Representative Luken. All proposed ash legislation considered thus far by Congress would eliminate the central issue over ash by exempting it from classification as a hazardous waste. In addition, these bills direct EPA to regulate all aspects of ash—from the handling of the ash at the incinerator to its transportation, storage treatment, and final disposal.

The Burdick and Luken bills lay out specific requirements for the design of ash monofills and codisposal landfills. These bills would also require EPA to develop criteria and testing procedures so that ash that met the criteria could be codisposed of with municipal solid waste in a Subtitle D landfill. Finally, they would require EPA to develop regulations for recycling and reuse of ash. The new regulations would be issued in one and one half to two years, and the landfill design requirements would become effective four years after enactment. Some interim requirements would also apply.

Landfill Design Requirements

Both bills include specific design requirements for ash monofills and codisposal landfills (see table 5–1 for summary; for comparison purposes, the table also includes design requirements for a Subtitle C hazardous waste landfill). Design requirements for monofills are similar under both bills: they must have one composite liner consisting of a

Table 5-1. Design Requirements for Ash Landfills Under Proposed Legislation

	Burdick S. 196	Luken H.R. 2162	Subtitle C Hazardous Waste Landfill
Monofills			
Total Number of liners	one composite	one composite	two or more (e.g., one composite or one composite plus one additional synthetic liner)
Liners must prevent migration of any constituents during facility operations (including post-closure period)	yes	yes	yes
Leachate collection system – above and between liners	——	——	yes
– location unspecified	yes	yes	——
Groundwater monitoring	yes	yes	yes
For monofills contraining fly ash only	one composite liner plus additional liner _or_ treatment	——	——
Codisposal landfills			
Total number of liners	two or more composite	two composite	two or more (e.g., one composite or one composite plus one additional synthetic liner)
Liners must prevent migration of any constituents during facility operations (including post-closure period)	yes	yes	yes
Leachate collection system – above and between liners	yes	yes	yes
Groundwater monitoring	yes	yes	yes
Alternate designs allowed in lieu of the above requirements	yes	yes	yes

bottom liner of natural materials and a top liner of synthetic materials, though only Senator Burdick's bill includes specific requirements for monofills containing only fly ash. Both bills would also allow the codisposal of ash and municipal solid waste under more stringent landfill designs.

Criteria and Testing Procedures

Both bills require EPA to develop criteria and testing procedures that would allow ash passing the criteria to be codisposed of with municipal solid waste in a Subtitle D facility provided that certain other requirements were met (see table 5–2). The standard for developing the criteria and testing procedures is stricter in Senator Burdick's bill than in Representative Luken's. In addition, S. 196 provides more detail than H.R. 2162 as to the specific factors that must be considered in the development of the criteria and testing procedures.

Codisposal Under Subtitle D

In addition to meeting new test criteria, both bills would require the facility to meet other requirements before the ash could be codisposed of with municipal solid waste under Subtitle D. Again, the requirements under Senator Burdick's bill are more rigorous than under Representative Luken's. For example, S. 196 requires at least one liner, a leachate collection system and groundwater monitoring while the additional requirements under H.R. 2162 are limited to dust emission controls and entry restrictions (see table 5–3).

Recycling and Reuse

The recycling and reuse provisions of the two laws essentially are identical. They both require the EPA Administrator to develop regulations for the recycling and reuse of ash "as may be necessary to protect human health and the environment." Requirements must include criteria and testing procedures that consider all potential pathways of exposure over the short or long term. In addition, ash must be treated before it is recycled or reused. The

only significant difference between the two bills is that Burdick's bill contains a legislative "hammer" which states that if EPA fails to develop recycling and reuse regulations within 36 months of enactment, no recycling and reuse will be allowed unless an extraction procedure test is applied to the treated ash and the resulting leachates do not exceed drinking water standards.

Interim Requirements

During the four–year period between enactment of legislation and the effective date of the regulations, some interim requirements would apply (see table 5–4). While Representative Luken's bill is more comprehensive than Senator Burdick's bill, the requirements under the latter are again more stringent.

Miscellaneous Provisons

Source Reduction. Both bills address source separation of waste to reduce toxicity of the ash. In the Senate bill, states or local governments would be required to revise their solid waste management plans to include provisions for source separation to reduce the toxicity of ash. The House bill would require EPA to issue guidelines within 18 months that would recommend methods for removing items or materials that could cause toxic ash or emissions.

Permits. Both bills would require states to implement permit programs for any facility managing ash.

Capacity. Only Luken's bill addresses ash management capacity. This bill would prohibit an incinerator from operating unless "adequate capacity is available." However, the proposed legislation does not specify over what period of time adequate capacity must be demonstrated.

Export of Ash. Senator Burdick's bill would place restrictions on the export of ash outside the United States. Exports would be allowed only if there was an international agreement and EPA prior approval.

Table 5-2. Criteria and Testing Procedures Under Proposed Legislation

	Burdick S. 196	Luken H.R. 2162
Basis for these provisions	to determine whether ash samples *may* pose a hazard to human health and the environment	to determine whether ash samples *will* pose a hazard to human health and the environment
Factors that must be considered:		
– potential contamination – specific ash properties – test leaching conditions – test pass/fail criteria	– to all potential pathways – heterogenous nature of ash – acidic and native – ash will fail test if >100 times drinking water standards	– only to groundwater or surface water —— —— ——
Deadline for prescribing criteria and testing procedures	——	within 24 months
Includes program to validate procedures by analysis of facility leachates	yes	——
No routine testing for facilities that meet design requirements	yes	——

Table 5-3. Proposed Legislation for Codisposal Under Subtitle D

	Burdick S. 196	Luken H.R. 2162
Ash can be codisposed of with municipal solid waste in Subtitle D landfills if:		
– criteria are met	yes	yes
– and –		
– additional requirements are met	– one liner – leachate collection system – groundwater monitoring – compliance with existing and revised Subtitle D criteria	– dust emission controls – landfill entry restrictions – compliance with existing and revised Subtitle D criteria
Additional requirements for fly ash and combined ash disposed of in Subtitle D facilities	– treatment – treated ash must not fail criteria	——

Table 5–4. Interim Requirements Under Proposed Legislation

Burdick S. 196	Luken H.R. 2162
Eighteen months after enactment, of the bill, ash landfills must have: – one liner – leachate collection – groundwater monitoring Variances allowed based on cost considerations	Sixty days after enactment of H.R. 2162, EPA must publish interim rules for the handling and transportation of ash and for operating controls at ash landfill including: – emission controls – run–on and run–off controls – surface water and access controls – closure and post–closure provisions Ten months after enactment of the bill, ash landfills must have: – one liner or leachate collection or both "whenever capacity is reasonably available" – groundwater monitoring Ash landfills must meet existing and revised Subtitle D criteria

Air Emissions. Senator Burdick's bill includes provisions for controlling air emissions from MWCs. The House bill does not. This difference may delay passage of any legislation.

Resolution Through Legislative Inaction

EPA expects to issue a final rule in 1989 that could reduce the maximum contaminant level (MCL) for lead in drinking water as much as ten-fold from 50 ppb to 5 ppb. In addition, the Agency has proposed cutting the MCL for cadmium in half. The permissible limits for lead and cadmium in the EP tox test, which determines whether or not a waste is toxic and therefore hazardous, are tied to these drinking water standards. Regardless of the exact level that is eventually set, any reduction will mean that at some later date the EP toxicity limits will also be lowered and then more ash, when tested, will exceed the EP toxicity limit (see test results presented in chapter 3). Because the TCLP uses the same regulatory limits, the changes to the

drinking water standards will have the same impact on both tests—unless EPA significantly alters the test.

Once the final drinking water standards are issued, EPA will adjust the EP tox or TCLP limits to reflect the new standards by issuing a separate rule —though whether the process will take one year, five years, or longer is unknown. Once the EP or TCLP limits are changed, a strong argument could be made for listing ash as a hazardous waste. Even should it not be listed, it can be expected that the public will increase demands for managing ash as a hazardous waste.

If all MWC ash is listed as a hazardous waste, many MWCs may have difficulty finding sufficient landfill space with which to dispose of their ash. Other MWCs may be able to find hazardous waste landfill space, but disposal and transportation costs will go up. The inevitable decrease in the amount of waste that will be incinerated is a result that will be welcomed by some and bemoaned by others.

REGULATORY SOLUTIONS

EPA has three regulatory options for MWC ash: ash can be managed as a hazardous waste under Subtitle C, as a nonhazardous waste under Subtitle D, or as a "special" waste under Subtitle D. Which one of these options EPA will ultimately use depends on actions by Congress and the courts. However, if Congress does not pass ash legislation and if EDF's lawsuits do not prevail, a regulatory resolution to the ash problem will be left entirely to EPA's discretion.

The first option, managing some MWC ash as a hazardous waste based on the results of the EP tox test, appears to be EPA's current position (though it has not enforced it). Yet because of the many problems with the EP tox test (see chapter 3), this will produce a poor ash management system. Replacing the EP with the TCLP will reduce but not eliminate some of the problems. Furthermore, ash *not* classified as hazardous would be regulated under Subtitle D. Even with the proposed revisions Subtitle D regulation may be insufficient for managing ash.

The second option, regulating ash under Subtitle D, raises several issues. EPA has stated that the recently proposed Subtitle D requirements, while geared toward the management of municipal solid waste, will be applicable to landfills receiving MWC ash (USEPA 1988c). Although these regulations are a vast improvement over the existing Subtitle D criteria, they do not address several important issues raised by the land disposal of ash. First, because no distinction is made between ash and municipal solid waste, the requirements do nothing to discourage codisposal. Yet, codisposal of MWC ash with municipal solid waste is exactly the type of situation the Subtitle C regulations were designed to prevent: a potentially hazardous waste disposed of in an environment that encourages leaching of significant quantities of hazardous constituents. While results frequently differ, some ash, as shown by numerous laboratory studies, surpasses the EP toxicity test limits for lead and cadmium. Furthermore, there is universal agreement that the acid pHs produced by decomposition of garbage favor the solubility of these metals. Congress may have exempted ash from Subtitle C for reasons not related to its potential toxicity, but by allowing ash to be codisposed with municipal solid waste, the criteria encourage an obviously more risky disposal practice.

Second, tests of MWC ash indicate that lead is the substance of greatest concern, so the presence of lead will probably drive risk assessments concerning MWC ash. Yet the design goal for landfills in the proposed criteria are based on risk assessments that consider only carcinogenic risks. Though EPA has classified lead as a probable carcinogen, there is no question that lead causes noncarcinogenic effects in humans at lower levels. Hence, the drinking water maximum contaminant level and thus the EP tox test limit for lead are based solely on these noncarcinogenic effects. Ignoring the noncarcinogenic effects in the assessment of risk from MWC ash may mean overlooking the most significant health risks.

The proposed criteria do address noncarcinogenic risks in the requirements for groundwater monitoring and corrective action, and these provisions will greatly reduce the chance that groundwater from a landfill contaminated by potentially harmful concentrations of lead or other toxic constituents will be consumed by the public. However, in light of the potential harm posed by the noncarcinogenic effects of lead in ash, a sensible regulatory program should not focus on detection and cleanup after the leachate has escaped the landfill.

Finally, the regulations provide for the selection of an alternative boundary as the compliance point where the design goal must be met. The alternative boundary, which can be up to 150 meters from the landfill boundary, allows "the mitigating effects of the surrounding soils and aquifer material to reduce the concentrations of the contaminants" (USEPA 1988a, p. 33354). This may be an acceptable practice around municipal solid waste landfills where toxic constituents in leachates should not be expected to exceed drinking water levels. In ash landfills, however, where concentrations of lead and possibly cadmium in the leachate might reach or exceed EP toxicity test levels, allowing the leachate to migrate far from the landfill before initiating the monitoring requirements may be an unsound practice. Moreover, the limit of 150 meters appears arbitrary. The proximity of water sources might be a better guide.

The third option—regulating ash as a special waste under Subtitle D—would produce the most sensible system.

There are two approaches EPA could take to pursue this option. In the absence of a specific mandate by Congress or the courts, EPA could take the initiative and either (1) develop regulations or else (2) issue guidance specifically for ash. EPA has authority under RCRA to develop regulations to manage ash differently (i.e., more stringently) than other Subtitle D wastes. Furthermore, EPA does not need special authority to make recommendations to owners or operators of incinerators on how to manage ash, and such information could be distributed by way of a guidance manual.

Despite this opportunity EPA is not likely to issue regulations or guidance without action by Congress. The Agency developed a draft guidance document in March of 1988, which provided recommended procedures for managing ash (EPA 1988a). While it was widely distributed, it was never officially released. Had it been, it might have served as a partial or interim solution to the ash problem.

SUMMARY

A decision in the lawsuits against two MWCs could clarify the ambiguous household waste exclusion. However, either possible legal outcome would not resolve all of the problems over ash management. Congress has two bills under consideration. Passage of either would be an important step toward a sensible solution. The two bills are similar in that both would exempt ash from hazardous waste regulations and require EPA to develop specific ash management requirements. Classifying ash as a hazardous waste is a possible outcome if Congress fails to enact legislation though how long

it would take before such a change was in place is unknown. Finally, EPA could move ahead without new legislation and issue regulations or guidance on ash but thus far has not pursued this option.

APPENDIX 5–A

Cost of Disposal Under Subtitle C and Subtitle D

In principle, the cost of management of hazardous wastes is higher than nonhazardous wastes because of the more rigorous requirements that owners or operators of these facilities must comply with under Subtitle C. Yet the impact to incinerator owners or operators of requiring ash to be disposed as a hazardous waste (rather than as a nonhazardous waste) is difficult to assess. A comparison of the price per ton to dispose of waste gives a general picture of the difference in disposal cost between the two Subtitles. In a survey of hazardous waste companies conducted by EPA (1988b), the price for land disposal of hazardous wastes in 1986–87 ranged from $75 to $250 per ton. A narrower range of $97 to $166 per ton reflects the average of all low–end and high–end prices quoted; $131 is the average of these two figures. A survey of tipping fees, prices charged to dispose of municipal solid waste in landfills in 1988, conducted by the National Solid Wastes Management Association (Pettit 1989), showed that these fees ranged from $4 to $132 per ton, with an average of $27 per ton.

This comparison between the disposal costs under the the two Subtitles does not consider that different types of hazardous wastes can be charged different prices. Nor does it consider the wide regional variation in municipal solid waste disposal costs, which is due in part to widely different state requirements. Finally, the revisions to federal requirements for municipal solid waste landfills will probably cause the average solid waste tipping fee to rise.

It is also important to note that direct comparisons of landfill fees do not take into account the costs of transporting the waste to the facility. Although the cost per mile to transport a hazardous waste is not much higher than for nonhazardous wastes, ash classified as hazardous would have to be transported much farther than nonhazardous ash because there are many fewer hazardous waste landfills than municipal solid waste landfills.

REFERENCES

Kovacs, W.L. 1988. "Can EPA Regulate Ash As a Hazardous Waste?" *Waste Age* (May) pp. 105–111.

Pettit, C.L. 1989. "Tip Fees Up More than 30% in Annual NSWMA Survey," *Waste Age* (March) pp. 101–106.

Thomas, L.M. 1988. Statement before the Subcommittee on Transportation, Tourism, and Hazardous Materials, Committee on Energy and Commerce, U.S. House of Representatives (April 13).

U.S. Environmental Protection Agency (EPA). 1988a. Draft Guidance: Municipal Waste Combustion Ash. EPA: Washington. (EPA/530–SW–88–006).

U.S. Environmental Protection Agency (EPA). 1988b. *1986–1987 Survey of Selected Firms in the Commercial Hazardous Waste Management Industry*, Final Report, prepared for the Office of Policy Analysis (March 31).

U.S. Environmental Protection Agency (EPA). 1988c. Proposed Revisions to Criteria for Solid Waste Disposal Landfills: A Summary (August 22).

6

Conclusions and Recommendations

In an ideal world, decisions regarding the management of environmental risks such as MWC ash would be based on complete knowledge of risks to human health and the environment. In the case of ash, decision makers would know what amounts of which toxic substances could be leached from ash, how much airborne ash could be deposited on the skin, inhaled, or ingested by workers and neighbors; and how much of the toxic substances could be taken up by exposed individuals and by ecosystems. Such information would make it possible to assess health and environmental risks and to calculate the relative efficacy of alternative disposal or treatment methods designed to reduce the risk. It would also be possible to calculate the costs of those methods. All of this information, combined with data on risks and costs of waste management methods—incineration, landfilling, recycling and others—would give decision makers the means to make informed comparisons among municipal solid waste management choices.

Such analyses are now impossible because none of the methods used to predict concentrations of toxic substances in leachates from ash has been shown to be sufficiently accurate to provide a reliable basis for risk assessment. In the absence of a validated method to predict concentrations, not even a crude risk–based system for managing ash is now feasible. However, efforts to improve risk assessment methods should be continued. Eventually, they may be incorporated into the ash management system.

LEGISLATIVE CHOICES AVAILABLE

The future of the MWC ash issue will depend on whether or not Congress addresses the issue. If Congress takes no action, confusion over the regulatory status of ash will persist. At some unknown future date, regulatory events that are already set in train could result in essentially all MWC ash being declared a hazardous waste (see Chapter 5), which would restrict incineration of municipal solid waste. Under a second scenario, Congress could pass new legislation that requires that ash be regulated as a nonhazardous waste. That legislation, however, would impose additional requirements to ensure that ash would be managed more stringently than other nonhazardous wastes. It also would favor continued operation of existing MWCs and, probably, the construction of new ones.

IMPLICATIONS OF EACH CHOICE

Regulation of MWC ash as a hazardous waste—either through congressional action or inaction—would have a profound effect on existing as well as planned incinerators. The stigma of classification of ash as a hazardous waste would, in itself, be a huge burden to all MWCs. Some existing MWCs might be forced to cease operation because of the higher cost and paucity of landfill space for hazardous waste disposal. Because approval of new incinerators usually requires a guarantee of adequate landfill space for ash disposal, new MWCs might have to construct new hazardous waste landfills in order to gain approval. Obtaining a permit for either an MWC or a hazardous waste landfill requires much time and effort and there is no guarantee of success; to couple these efforts would make licensing of new MWCs extremely difficult.

On the other hand, the proposed legislation that would regulate ash as a nonhazardous waste with specific requirements for ash management would preserve incineration as a solid waste management option. Such action is not likely to cause a boom in MWC construction because cost alone is likely to limit the number of MWCs no matter what the regulatory program is. Such legislation would, however, require the collection of more information about ash and possible exposures to it. Should that information reveal unacceptable concentrations of toxic substances in leachates, more stringent disposal methods could be required, or in the extreme, MWCs could be closed.

Until Congress enacts legislation or the regulatory process classifies ash as a hazardous waste, EPA is unlikely to issue specific ash regulations or guidance documents. The current situation, in which EPA awaits a congressional decision, provides little opportunity for any organization other than Congress to make policy.

Our recommendation begins with a call for Congress to enact legislation to resolve the regulatory controversy over ash. Next, we recommend actions by EPA that include the development of technical guidance and appropriate tests, and the collection of field data. Then, we present specific recommendations for managing ash under Subtitle D that rely on design specifications for ash landfills and not on results of tests to estimate the risks that may be posed by ash. We also recommend that not all ash be classified as a hazardous waste. In the long term, ash management should have some basis in risk. We suggest an approach for integrating tests into the ash management system. Our recommendations conclude with suggestions for further research.

RECOMMENDATIONS FOR CONGRESSIONAL ACTION

Congress needs to enact legislation that outlines a system for managing ash. Specifically, Congress should clarify the household waste exclusion which would determine whether or not ash can be classified as a hazardous waste. Legislation introduced to date would exclude MWC ash from regulation under Subtitle C and manage it under Subtitle D. Additional requirements specific to ash management would also be required.

Enacting ash legislation is the single most important action that could be taken to improve the management of MWC ash. Its absence burdens local and state governments that must make decisions about incinerators and allows some unacceptable ash management practices to persist. Without legislation, EPA almost certainly will not address the ash issue, and a decision in individual court cases, whatever the outcome, will not resolve underlying issues.

Although Congressional action is needed to resolve the central controversy over ash, the specificity in some provisions of proposed ash legislation may prove deleterious to managing ash in the long term. For example, one bill requires that tests be developed and spells out pass/fail levels for leachates. This level of detail in the proposed legislation must be disquieting to those who believe that technical competence and expertise resides in the regulatory agencies and not in Congress. Although Congress makes no claims for such competence and expertise, it appears to be set on specifying how ash is to be managed in some detail. This move toward "micromanagement" might be evidence of a general trend in congressional action, of dissatisfaction with EPA's inability to act in the case of MWC ash, or of concern that the agency might act more or less stringently than Congress thinks appropriate.

RECOMMENDATIONS FOR EPA

EPA has asked Congress to clarify the household waste exclusion and has since made little progress toward regulating ash. While waiting for congressional action, the agency could provide guidance about managing ash, pursue test development, and collect field data.

With or without new ash legislation by Congress, EPA could provide technical guidance for managing MWC ash. Such action by the agency would provide needed leadership to state and local governments. This guidance could be issued immediately and would be especially welcomed if actions by Congress are further delayed.

EPA should develop and validate appropriate laboratory methods to characterize ash. EPA should strive to improve the existing test as well as validate other tests that could be suitable for characterizing ash. Better tests are needed if ash management is to have any basis in risk, and EPA is the most appropriate place for their development. Some of the problems with the EP tox test will be eliminated when the TCLP is finally issued. However, some issues, such as the selection of the appropriate extraction fluid when testing multiple samples from a single MWC, still remain. A test that mimics reactions in monofills, such as the solid waste leaching procedure, should be validated and incorporated into the regulatory system. Finally, EPA must improve methods for obtaining representative samples.

More information about field leachates should be gathered. Where feasible, EPA should collect field leachate data regularly. The resulting information would assist EPA in assessing whether monofills with single composite liners provide adequate protection of human health and the environment. If these assessments indicate that monofills do not provide adequate protection, more stringent controls could be added or some form of treatment required. The requirement for monofills could also be reduced if new information shows that fewer controls are adequate to protect human health and the environment. Field data would also be used to provide information for the validation of tests.

ASH MANAGEMENT IN THE SHORT TERM

Until more information is gathered about actual field leachates, ash should be disposed of in monofills that have at least two liners, a leachate collection system, and groundwater monitoring. In addition, monofills need covers to reduce the amount of precipitation infiltrating into the facility. Monofills (either a single composite liner or two synthetic liners) are the most sensible short–term management approach for two reasons. First, laboratory test data show that acidic conditions, which can be caused by municipal solid waste, promote the leaching of toxic constituents—particularly lead and cadmium—from the ash. Although field data do not strongly support this finding, such data are sparse and there are no results from long–term field studies (i.e., longer than 10 years). Until more field data are collected and the potential for leaching from monofills and codisposal landfills is better assessed, monofills are the preferred option. Second, ash disposed of in monofills is recoverable, unlike that which is mingled with municipal solid waste. Thus, if containment is insufficiently stringent because leachates are more of a problem than they now appear, the monofilled ash could be excavated and treated, treated *in situ*, or excavated and placed in better designed facilities with no harm done. On a more positive note, if markets were found for ash reuse, the ash could be easily recovered from monofills.

Our recommendation for monofill disposal may be difficult for some owners/operators who currently rely on codisposal of ash. Rather than requiring these MWC owners/operators to site a new monofill, we suggest that, when appropriate, EPA allow them to construct a new cell (a lateral, not vertical expansion) at an existing municipal solid waste landfill.

Codisposal of ash and municipal solid waste may be allowed but only under more stringent disposal requirements. We recommend that these facilities include at least an additional synthetic liner. But because the overall advantages of monofills are greater, we prefer monofills to codisposal landfills.

Certain types of codisposal might be allowed in landfills with only a single composite liner. Noncombustible items can be removed from municipal solid waste before it is put into an incinerator. These items such as ferrous metal, bricks, and stones should not affect pH and should contribute little to toxicity. EPA could address the question of whether codisposal of those items with ash should be regulated differently from codisposal with municipal solid waste that contains organic material.

Our interim recommendations are similar to those of the Association of State and Territorial Solid Waste Management Officials (ASTSWMO). Expressing confidence that MWCs have a role to play in municipal solid waste management, along with recycling and landfilling, ASTSWMO recommends that MWC ash be regulated as a special waste under Subtitle D and that it be disposed of in lined monofills with no dependence on tests for classification.

The owners/operators of MWCs should not be required to keep fly ash and bottom ash separate. Most MWCs combine fly ash and bottom ash during the incineration process and mandating that these facilities keep fly ash and bottom ash separate would require changes in facility processes. In addition, there are some advantages to combining fly ash and bottom ash. Facilities that have scrubbers to control air emissions add lime, which is alkaline, to flue gases which ends up in the fly ash. When combined with the bottom ash, the basic fly ash mixture helps the ash to set up like concrete, thus reducing the likelihood that toxic constituents will leach. Combining fly ash with bottom ash within the incinerator may also help keep fly ash particles from becoming airborne, thereby reducing potential exposures to workers and the public from ash dust emissions. However, bottom ash alone, because it generally has lower metals concentration than fly ash, may have a greater potential for reuse than combined ash. Thus, one drawback to combining fly

and bottom ash may be that the reuse of bottom ash would be reduced.

MANAGING ASH AS A HAZARDOUS WASTE

Not all ash should not be managed as a hazardous waste. At some time in the future (as discussed in Chapter 2), the regulatory limits for lead and cadmium in the EP tox test or its replacement will probably be made more stringent. When that happens, ash leachates from many more ash samples, which are now comfortably below the current limit, will exceed the regulatory limit. With these changes, there could be a strong argument made for listing all ash as a hazardous waste. On the surface this appears to be an environmentally sound solution to the ash problem, but it may not be. If all ash is managed identically, whether it is managed as a hazardous waste or in monofills as a nonhazardous waste, there is little incentive to reuse the ash or to reduce ash toxicity through treatment or source reduction measures. These measures would be encouraged if appropriate tests for characterizing ash and more options for managing ash were available.

ASH MANAGEMENT IN THE LONG TERM

In the long term, the management of MWC ash should be based, at least in part, on an assessment of risk through testing. The short–term recommendations for identical management of all ash focus narrowly on use of monofills for containing ash. That approach to ash management, although an improvement over the current situation, provides little incentive to reduce the toxicity of the ash itself. However, the move to incorporate tests into the ash management system as data become available would encourage source reduction and treatment efforts. If those efforts reduce concentrations of toxic substances in ash, incinerator owners/operators that use them could adopt less stringent (and less costly) containment methods.

Two types of tests would be appropriate for inclusion in the ash management program. The first test would mimic long–term conditions at monofills and would probably use water (or rain water) as the extraction fluid. Ash that exceeds certain criteria in laboratory tests, however defined, could be made subject to more stringent controls, such as treatment or placement of the ash in a more secure landfill. Ash that meets the criteria could be disposed of in monofills, or it could be analyzed with a second test. The second test would mimic long–term conditions at codisposal landfills. Although some modifications may be necessary, EPA's replacement for the EP tox test, the TCLP, is an appropriate starting place for the development of a codisposal test. Ash meeting the criteria for both tests could be codisposed of in a municipal solid waste landfill.

The availability of reliable tests would not mean that MWC owners/operators would be able to change their ash management plans based on tests of ash on a daily, weekly, or some other routine time basis, obtain the test results, and then decide how to dispose of the ash. In fact, the availability of tests will be most valuable at the time that new landfill space is being obtained because owners/operators must contract for or otherwise secure landfill space for years into the future.

RECOMMENDATIONS FOR RESEARCH

Municipal incinerator ash has been recognized as a problem for only a few years, and many questions about it remain unanswered. Some of those questions can be answered through research, and the answers would improve ash management.

Reconsider the validity of factors such as attenuation and dilution in constructing tests. Currently, the EP tox test and its replacement, the TCLP, rely on an attenuation/dilution factor of 100 (above drinking water standards) for determining the allowable levels of toxic constituents in test leachates. This factor has little basis in actual field measurements or theory, and improvements to it should be considered. Some specific questions follow:

- How much are lead and other metals attenuated in clay liners or liners of natural materials? Should this attenuation be considered in

setting the acceptable limits for passing the test? Currently, the EP tox test assumes that no attenuation occurs.

- Should different attenuation/dilution factors, either higher or lower, be set for different substances?

- Should a single attenuation/dilution factor be used for all sites, or should site–specific models be developed to allow for different soil conditions, rainfall amounts, and other factors?

Continue to explore the potential of source reduction to reduce leachate toxicity. Source reduction is a broad term that describes many different activities including product modification and substitution by manufacturers, recycling by consumers, and presorting at the MWC to remove potentially toxic items. All these methods could reduce the toxicity of leachates from ash.

Two studies have examined the sources of lead and cadmium found in municipal solid waste. However, little is known about which materials in municipal solid waste contribute to the *leachable* lead and cadmium that are detected in tests. Some items may be major contributors of toxic constituents in municipal solid waste but contribute little or nothing to the concentration of those constituents in the leachate. Additional studies are needed to pinpoint the items that are "precursors" to the toxic constituents in ash leachates, followed by analyses of the feasibility and effects of removing such items from the waste stream. Initially, source reduction activities should focus on lead and cadmium because those two metals are the focus of current concern in ash management.

Gather more data on the behavior of highly alkaline ash and its effect on clay liners. MWCs that have scrubbers add lime to control air emissions. This lime ends up in the fly ash and, as a result, the ash can be very alkaline. Lead, unlike most other metals, is soluble in alkaline solutions and therefore conditions are created that encourage it to leach. This has been demonstrated in limited batch tests, but there are few, if any field data. There is also evidence that clay liners break down when exposed to highly alkaline leachates. More data need to be collected about the leachability of lead from alkaline ash and the effect of alkaline leachates on clay liners. If necessary, alternative landfill designs and treatment methods should be explored.

Find out more about the natural solidification properties of ash. Some ash, particularly ash that has lime added to it, tends to set up like concrete. This property of ash, which may greatly affect leaching, has been given little attention by regulatory agencies. Some questions that need to be addressed are:

- Under what conditions does ash solidfy? How does solidification affect solubility of toxic substances?

- How durable is solidified ash? How permeable is this type of ash to water? Does leachability change over time?

- What are the concentrations of toxic substances in leachates and runoffs?

- Should solidification of ash be considered a treatment method?

SUMMARY AND CONCLUSIONS

Decisions on how to appropriately manage MWC ash remain clouded with confusion and controversy. Federal laws and regulations that mandate how ash is to be managed are unclear, and tests for determining whether it is a hazardous waste produce ambiguous results. This situation has served both to restrict the use of incineration and to delay federal action to improve ash management. Although decisions about solid waste management, including the acceptability of incinerators, should be made locally, the federal government should not eliminate or restrict any solid waste management option—such as incineration—unless it has evidence that it presents an unreasonable or unman-

ageable risk to human health or the environment. And although MWC ash contains toxic chemicals that, under certain worst-case conditions, have the potential to contaminate groundwater supplies, pollution problems from ash can be prevented by proper management.

The most important action to resolve the controversy over ash is for Congress to enact legislation that outlines a system for managing ash. That legislation should call for disposing of ash in monofills with no reliance on tests. However, EPA should continue to develop better tests and, ultimately, tests should be used to make management decisions about ash. We recommend that the tests for MWC ash be designed to predict how ash will behave under likely disposal conditions in addition to those that mimic worst-case conditions, even though this approach departs from the current regulatory structure under RCRA. Managing ash with some basis in risk assessment has the advantage of encouraging treatment and source reduction. Because better information about ash will improve its management, we also describe areas for further research. However, by no means should efforts to resolve the ash controversy wait until these research questions are answered. Sufficient information is already available to put a workable system into place, and resolution of the problem is long overdue.

The controversy surrounding ash brings to light some of the inherent problems with our current system for managing this nation's wastes. Dividing all wastes into but two categories—hazardous and nonhazardous—for purposes of regulation provides few options for those wastes such as MWC ash which do not fit clearly into either. Another problem is that because the system focuses on how to manage the waste after it is produced, there are few incentives to use innovative treatment methods or source reduction measures. A risk-based system would improve the management of ash and other wastes. That system, however, must await the development of appropriate tests and policies.

Workshop Participants

Ms. Kerry Callahan
Association of State and Territorial
  Solid Waste Management Officials

Dr. Richard Denison
Environmental Defense Fund

Ms. Madeline Grulich
Virginia Department of Waste
Management

Mr. Floyd Hasselriis
Gershman, Brickner & Bratton, Inc.

Ms. Elaine Koerner
Resources for the Future

Mr. James E. McCarthy
Congressional Research Service

Mr. Frank McManus
Resource Recovery Report

Mr. James Pittman
Hazardous & Solid Waste Management
 Administration
Maryland Department of the Environment

Dr. Paul R. Portney
Resources for the Future

Dr. Walter M. Shaub
Coalition on Resource Recovery
 and the Environment
U.S. Conference of Mayors

Dr. Doreen Sterling
Office of Solid Waste
U.S. Environmental Protection Agency

Mr. David B. Sussman
Ogden Projects, Inc.

Mr. Frank Sweeney
Subcommittee on Commerce,
 Consumer Protection and Competitiveness
U.S. House of Representatives

Mr. Carlton Wiles
Office of Research & Development
U.S. Environmental Protection Agency

Mr. Michael Winka
Division of Solid Waste Management
New Jersey Department of Environmental
 Protection

Report Reviewers

Mr. William C. Child
Division of Land Pollution Control
Illinois Environmental Protection Agency

Dr. Richard Denison
Environmental Defense Fund

Dr. T. Taylor Eighmy
Department of Civil Engineering
University of New Hampshire

Mr. Ulysses G. Ford, III
Solid Waste Management Department
City of Houston

Mr. Floyd Hasselriis
Gershman, Brickner & Bratton, inc.

Dr. Howard Levenson
Office of Technology Assessment

Ms. Sylvia K. Lowrance
Office of Solid Waste
U.S. Environmental Protection Agency

Mr. Norman H. Nosenchuck
Division of Solid & Hazardous Waste
New York Department of Environmental
 Conservation

Mr. Ted Siegler
DSM Environmental Services, Inc.

Dr. Walter M. Shaub
Coalition on Resource Recovery and the Environment
U.S. Conference of Mayors

Mr. Claude E. Shinn
Oregon Department of Environmental Quality

Mr. Edmund J. Skernolis
Waste Management, Inc.

Mr. David B. Sussman
Ogden Projects, Inc.

Ms. Marcia Williams
Browning–Ferris Industries

ADDENDUM

Just before this report was to be printed, a decision was issued in the United States District Court, Southern District of New York, in a lawsuit against a municipal waste incinerator (*Environmental Defense Fund, Inc. v. Wheelabrator Technologies Inc. and Westchester RESCO Company, L.P.*, 88 Civ. 0560 (CSH), November 21, 1989). The owner of the incinerator had been charged with violation of federal hazardous waste law (Subtitle C of RCRA) because the facility produced ash that would be classified as a hazardous waste according to results of the EP toxicity test. The defendants argued that the facility was exempt from Subtitle C because of the household waste exclusion (see Chapter 2, p. 15 for further discussion). The court issued a memorandum opinion and order which, assuming it is upheld at higher courts, resolves the legal ambiguities over the regulatory status of ash.

The judge in this case ruled that the household waste exclusion does exempt the facility from Subtitle C of RCRA. In other words, ash from most municipal waste incinerators cannot be classified as a hazardous waste, regardless of its toxicity. The judge reasoned that this was Congress' intent, based on the legislative history of the 1984 RCRA amendments.

One issue in this particular case remains unresolved. For the household waste exclusion to apply, the facility must accept only municipal solid waste. The plaintiff charges that the facility may be accepting hazardous wastes from small quantity generators or from illegal generators. The court has ordered that the discovery process go forward to obtain information about this point. Thus, it still remains possible that the court may find that the facility is not exempt from Subtitle C requirements.

Congress intended to encourage incineration by including the household waste exclusion in the law when it amended RCRA in 1984. Yet it could not have anticipated the controversy over the toxicity of ash that would arise in the years to follow. The implication of this court decision is that, at least under federal law, ash from municipal waste incinerators can be managed as a Subtitle D nonhazardous waste (or under more stringent state laws, where such laws exist). While some owners and operators of incinerators are now "off the hook" and need not worry that their ash may be declared hazardous, the issues surrounding the appropriate management of ash are far from being resolved. Much ash still fails the EP toxicity test, even if it cannot be subject to hazardous waste requirements. And although we do not advocate managing all ash as a hazardous waste, there are few who would argue that the existing nonhazardous waste requirements are adequate for managing ash. As the legal issues are resolved, the pressure for Congress to act may be lessened. We believe now, more than ever, that federal leadership is crucial in developing information about ash and ensuring that ash is managed properly if we are to rely on incineration in the management of municipal solid waste.

For Product Safety Concerns and Information please contact our EU
representative GPSR@taylorandfrancis.com
Taylor & Francis Verlag GmbH, Kaufingerstraße 24, 80331 München, Germany

www.ingramcontent.com/pod-product-compliance
Lightning Source LLC
Chambersburg PA
CBHW082107210326
41599CB00033B/6623